Post-Global Network
and Everyday Life

Steve Jones
General Editor

Vol. 60

PETER LANG
New York • Washington, D.C./Baltimore • Bern
Frankfurt • Berlin • Brussels • Vienna • Oxford

Post-Global Network and Everyday Life

EDITED BY MARINA LEVINA AND GRANT KIEN

PETER LANG
New York • Washington, D.C./Baltimore • Bern
Frankfurt • Berlin • Brussels • Vienna • Oxford

Library of Congress Cataloging-in-Publication Data

Post-global network and everyday life / edited by Marina Levina, Grant Kien.
p. cm.— (Digital formations ; vol. 60)
Includes bibliographical references and index.
1. Internet—Social aspects. 2. Globalization.
I. Levina, Marina. II. Kien, Grant.
HM851.P669   303.48'33—dc22   2010002719
ISBN 978-1-4331-0699-6 (hardcover)
ISBN 978-1-4331-0698-9 (paperback)
ISSN 1526-3169

Bibliographic information published by **Die Deutsche Nationalbibliothek**.
Die Deutsche Nationalbibliothek lists this publication in the "Deutsche
Nationalbibliografie"; detailed bibliographic data is available
on the Internet at http://dnb.d-nb.de/.

© 2010 Peter Lang Publishing, Inc., New York
29 Broadway, 18th floor, New York, NY 10006
www.peterlang.com

Table of Contents

Acknowledgments

The editors would like to thank Steve Jones for his excitement and support for this book project, and all the contributing authors for your commitment to outstanding scholarly work.

Control and Fear in Post-Global Network

MARINA LEVINA & GRANT KIEN

Our present millennium began with the Y2K scare, inducing widespread anxiety about a potential collapse of the hybrid digital and economic global network so carefully constructed throughout the latter part of the 20th century. As Charlie Gere (2002) points out, Y2K brought with it a sudden popular awareness of how much we had come to depend on this global system for our survival, and the extent to which the operational concept of a global digital and economic network had come to displace traditional linear notions of hierarchy, authority and power. The following year, the 9/11 terrorist attacks on the World Trade Center in New York City and the Pentagon in Washington, DC introduced a new concern into our milieu of cultural narratives, serving as a coming out for an enemy that was said to be just as globally-networked and invisible as the system we depend upon for our daily sustenance—and just as unpredictably unmanageable.

Where Y2K had illustrated the potential for benign technological catastrophe, 9/11 showed the world that the global network was also susceptible to intentional human acts of subversion and sabotage. A rash of science fiction media narratives such as *Battlestar Galactica* (2004), *I, Robot* (2004), *Terminator 3* (2003), and others, reflected a popular paranoia about the vulnerability of relying on the network.[1] Unable to pin down a singular, containable enemy in the traditional sense, the descriptor 'network' became conceptually configured as a singularized enemy, something to be "brought down." But here we have an irony brought on by a conflict in signification. The post-global network both overtly denotes danger, while at the same time more subtly connoting freedom.[2]

Even while contending with the newly-discovered vulnerability of being networked and actively fighting "the terrorist network," new "social" network technologies have continued to proliferate, enabled by Web 2.0 programs such as Facebook and YouTube, MMORGS like World of Warcraft, convergence across and between various media platforms, and ever increasing levels of global economic and cultural networks. In spite of its known vulnerabilities and dangers, the masses have continued to cavort with the enemy. We fear it and fight it, even while we crave it and embrace it as a positive force in our lives. In spite of this tension (perhaps even assisted by it), network has come into its own as a state of mind and a way of life—in sum, a cultural norm. From Barackobama.com to GoogleHealth to Twitter, a multitude of cultural and individual activities have become seamlessly reconfigured and woven into the ever-shifting spatial ambiguity and temporal presence of our hybrid global network. As a result, it is no longer fitting to examine the network as an external force, but rather as a somewhat banal aspect of our everyday environment. We not only performatively inhabit and enact moment to moment, we have also embraced the reality of being globally networked as a prevailing logic in our everyday experiences. Network, therefore, should not simply be conceptualized as a singularity or a technological entity, but rather as an always-already amorphous condition of life itself. As crucial as the chips and wires that comprise and connect the appliances we use, we everyday users are a fundamental component of network. So complete is its weaving into global economics, communications and our everyday lives—indeed, even our ontology[3]—that we here claim to be living in a state of post-global network.

This book project was originally inspired by a conversation about the zombie apocalypse film *28 Weeks Later* (2007). A sequel to the popular *28 Days Later* (2003), the movie portrays London twenty-eight weeks after the "rage" virus[4] that destroyed England by turning the majority of the population into the "infected"—aka zombies—was effectively contained and eradicated. In the sequel, US-led military forces reopen London for repopulation. Forming a "Green Zone" inside the city (a heavily surveilled, guarded, and regulated territory), the military allows limited numbers of refugees to return to their hometown. However, repopulation and re-containment fail as the virus reappears in the population. In a desperate attempt to reestablish dominance, the military gives up on selective elimination of the "infected" and instead issues the command: "Abandon selective targeting. Shoot everything. Targets are now free....We've lost control." This particular scene in which the enemy is no longer discernable from the general populace, and the chosen military is the solution to the problem, highlighted one of the key questions that prompted this book project: What is control in the context of network?

The phrase "We've lost control" represents an anxiety we consider to be discernable at the heart of efforts to contain and control threats in the post-9/11, post-global network era. *28 Weeks Later* exemplifies the futility of modern warfare in post-global, viral conditions of terrorism, epidemics, and cyberwars. The 'enemy' is indistinguishable from the general populace, and hence the amorphous multitude itself becomes the target. From zombies to zombie-bots, the rise of network society, and with it a system of power relations necessitated by a world in which globalization is a fact of life, the film illustrates an overarching problematic of what it means to fight for control of a ubiquitous, self-sustaining yet indefinable entity. We argue that this problematic is tethered to a yet to be fully explored quality of the network as an active, seemingly alive post-human (i.e., Human-machine hybrid) entity that, by its very nature, complicates a hierarchical order of power and knowledge.

David Singh Grewal (2008) argues that, as a non-linear power relation, working through decentralized relations of sociability, the network operates through regulations of standards as opposed to the enforcement of a sovereign will. This does not mean that network is democratic. As actor-network theorist Bruno Latour (1988) pointed out more than two decades ago, all relations in network are demonstrated lines of force. Network functions as a diffuse system of control and regulation operating through a multitude of nodes. Without institutions to contain it and give it form, network becomes difficult to fix, manage and control. Moreover, network is more than the accumulation of individual bodies. It is more like what Terranova (2004) refers to as a system of multitude. In this system, she writes:

> …you can observe and kill an individual entity, anatomize it, and you still won't find out what it is that will make it act in a certain way once it acts as an element within a population open to flows. You can collect as much data as you want about individual users, but this won't give you the dynamic of the overall network (p. 104).

This system invites, she argues, "the abstract machine of soft control," that facilitates growth and function of a multitude. Building on Hardt and Negri's (2000) autonomista discourse, she explores the notion of multitude as a potential for political engagement in the network. Multitude invites a different type of control from hierarchical systems and therefore different power alliances and relations. These are made even more slippery in the post-global context. Much like network has become a cultural norm, so too has globalization irrevocably changed the world order (in the sense of a one-world government). And in the context of Hardt and Negri's *Empire*, the post-global network necessitated the emergence of new sovereignty aimed at controlling multitude. With principles of network organization at its heart, a post-global sovereignty in the context of statehood "does not annex or destroy the other powers it faces but on the contrary opens itself to them, and thus through the constitutive network of powers and counterpowers the entire sovereign body is continually reformed." (Hardt and Negri 2000, p. 166) Referring to 'principles of network' as an organizational concept emphasizes not just the macro-political, but also strong socio-cultural dimensions of network. It is fitting to thus briefly turn to sociology to understand the social nature of this machinic assembly.

We are living in what has come to be popularly referred to as "network society." (see Castells 2000, 2001, 2006; Craven and Wellman, 1973; van Dijk, 1999; and Wellman 1979, 1988, 2001) To generalize the various similarities in seminal writings employing the term, there is agreement that network plays a fundamental role in the shaping and stabilizing of society. Implicit in this acknowledgement is a post-information age assumption that communication provides network with materiality, conceptual and ideological stability, and process in terms of procedure and everyday enactment. In sum, communication constructs the infrastructural "space of flows," (Castells, 2000) is the product that passes through those spaces, and provides us and our machines with the performative scripts we rely on for stability and function in everyday experiences. Network society theorists have expressed various opinions on the issue of whether the sum of network is greater than its parts—i.e., which is the basic unit of network, the array or the individuals in it? In contrast, actor-network theorists worked forward without addressing this question of fundamental premise, exploring the nature of network as a socio-technical hybrid to great success.

There is much that can be learned from actor-network theorists about the dynamics and ramifications of network as a social arena.[5] For our purpose here, it will suffice to limit our brief encounter with this body of work to the concept of control, which, in Actor-Network Theory (ANT), is said to be manifest through translation. By speaking on behalf of all other entities comprising an actor-network, the translating voice takes on the appearance of rising above and dominating the network. This is not just a simple act of standing on the shoulders of others. By enlisting other actants as supporting entities, the dominating voice "translates, explains, understands, controls, buys, decides, convinces, and makes them work." (Latour, 1988, p. 172) Rather than the center/periphery model of power relationship we are accustomed to discussing in a traditional linear model of organization, the result in networks is something known as a "center of translation." (Law, 1992, p. 388) These centers are relational effects, generated by conditions and material that can also result in their dispersal upon cessation or removal of the conditions or material which gives rise to them. Thus, for ANT, control in network is semiotic in nature, and understanding control and power entails examining semiotically the relations of materials and processes of communication that create centers of translation.

These power dynamics make the network subject to takeovers and holdovers; an open, expansive entity of constant resistance as opposed to the modernist idea of constant containment and control. The behavior of nodes in the network is dependent on the network's topology; in those terms, control in the network only functions as long as topologies are not affected, or infected, by other life. This is what Galloway and Thacker (2007) call an exploit; an asymmetric conflict in which networked actors struggle over centralized powers. They write that "this type of asymmetric intervention, a political form bread into existence as the negative likeness of its antagonist, is the inspiration for the concept of 'the exploit,' a resonant flaw designed to resist, threaten, and ultimately desert the dominant political paradigm." (p.21) Moreover, constant movement of information and nodes inherent in network encourages volatile spaces, random relationships, and zombie emergences. The zombie viral threat in *28 Weeks Later* illustrates metaphorically real world issues such as the Ebola virus, swine flu, and terrorist attacks both virtually and materially focused on disrupting or subverting informational flows (i.e., hijacking the media system with violent images). These all exemplify an exploit—a topological hole that threatens the characteristic of control in the network. The threat is uncontrollable precisely because it is asymmetrical to the original mode of domination: not counter-information collecting efforts, not a military force resisting another military force, and not an organized collective trying to destroy another collective. They are akin to computer viruses and as such "they exploit the normal functioning of their host systems to produce more copies of themselves. Viruses are life exploiting life." (pg. 83) In terms of power and control, this is conflictual but not in the way we've traditionally thought of conflict as dialectical struggle.

Virilio (2000) wrote, "Tomorrow's war will be *globalitarian*, in which…the qualitative will be of greater importance than geophysical scale or population size." (p. 144) The tomorrow he was referring to in 2000 became reality in 2001 with the 9/11 attack. Seeming to parallel the actor-network theorists' conclusion that struggles in networks are fundamentally about semiotics, Virilio (2002) described this new form of warfare as a battle over the world's images, claiming that the "optically correct" succeeds the "politically correct."(p. 31) In this situation, democracy becomes the free circulation of aesthetic representations, and the freedom to choose which representations one wishes to consume. However, the unceasing growth of the global system brings its own dangers. Every addition to the network adds more weight and content such that the system could accidentally collapse under its own weight, much like a building frame that is unable to support its roof. More ominously related to issues of power and control, the steady increase in mass makes the system both more open to potential threats (as every addition creates more opportunities for potential exploits as described above), and makes the stability of the system easier to disrupt (as its supports become strained under the burden of the load). In this condition, disruption in the system is never benign. Rather, it is individually devastating at best and systemically—even globally—catastrophic at worst. Thus expressions of agency appear terroristic in nature. After the 9/11 attacks, Baudrillard (2002) described the global network in this condition to be prone to "terroristic situational transfer," (p. 9) in which control of the system is wrested through an intentional catastrophic intervention. In this complicated milieu, users are inseparable from the technology of the network, creating a hybrid within which it is nearly impossible to distinguish enemy from friend. As Baudrillard (1988) put it, "There are no more individuals, but only potential mutants." (p. 51) But then, one might ask, is my own life experience not unique, at least to myself if not someone else? Is this not impossible for me to experience someone else's life, and this being the case, it must follow that even though we may be systematically assimilated and appended, individuals from their own perception continue to experience everyday life in network as authentically unique as ever.

To conceptualize assertions of control in the network, then, we have to make allowances for that part of the network that are animated and acting without being human. As a result, to "shoot everything" is not a viable option. Instead what is needed is a systematic engagement with the nature of network as we experience it in the post-global world. The chapters in this collection do not aim to reestablish control, but rather to get a handle on "lived" aspects of being networked. As such they explore everyday life in the condition of global network.

Jack Bratich uses the 9/11 truth movement as a case study in his chapter "When Collective Intelligence Agencies Collide" to discuss how suspicious subjects and identities are produced in the network. He examines how the collective intelligence of the network is shaped through its various institutional histories and how the distinction between "truth" and "conspiracy" is subject to political warfare conducted in the network. In his chapter "Intellectual Inquiry in the Age of The Efficient Network", James Salvo further pursues the question of knowledge formation in the network. Drawing on Walter Benjamin's work, he investigates how the seemingly infinite library the network offers knowledge-seekers shapes and alters the nature of intellectual inquiry. He argues that the unbounded nature of network library not only has implications for how knowledge is pursued, but it also shapes actual topologies of our everyday intellectual experiences. Continuing the investigation of knowledge and network, in "Travel in the Network", Ulrike Gretzel examines how the nature of travel has been reshaped by network infinities, palpably demonstrated through mobile technologies such as the iPhone. She argues that mobile technologies enable travel without ever leaving home; a network juxtaposition that affects the nature of an encounter with the "exotic" culture. While network can be experienced as though infinite, it is important to acknowledge that membership in any particular network is often restricted and unequally distributed.

Moving to politics of identity, Joy Pierce critically engages with the notion of the digital divide and how it shapes membership in the network in her chapter "Membership in the Network." She conducts a systematic study of how underrepresented populations are further excluded from the network through the development and design of the products that purport inclusion. In "Voicing and Placement in Online Networks", Radhika Gajjala and Anca Birzescu further this important topic in their study of race, ethnicity, and socio-economic globalization in cyberspace. They investigate how social network systems, particularly kiva.org, shape the issue of voice, subjectivity, and agency. Moving from the personal to the organizational, Michael Giardina provides a lively overview of online political organizing through a historical evolution in his work "From Howard Dean to Barack Obama". He argues that these examples demonstrate the evolution of networked political articulations and potential for group politics in the network. In another study of what social organization means for the network, Marina Levina's chapter, "Health 2.0 and Managing 'Dividual' Care in the Network", investigates the emergence of the Health 2.0 movement and its efforts to revolutionize health care access and provisions through online social organization. She examines what is at stake in Health 2.0's equivocation of individuals' health to the health of the network and the assumption that access to health information technologies is unproblematically linked to better health for all.

Focusing more on the nature of network and its implications for content, Sean Smith explores organization of sports and particularly how professional sport and network technologies organize and manufacture professional sports contexts in his chapter "Sport in the Wires". Using baseball as a case study and mindful of its political economy, he examines how sports are played in the wires. With network TV as a prototypical case study, James Hay explicates network as a system of governance in his chapter "Television as Everyday Network of Government". Crafting an overview of the history and evolution of network as a political formation, Hay brings the discussion into the present moment and illustrates how, by reinventing itself in the context of convergence and network society, network television has revamped its definition of itself as an arbiter of the public good. Finally, Grant Kien discusses the appropriation of labor in post-global network in his chapter "Privacy As Work," illustrating a new form of labor appropriation that has been enabled by the growth of social networking and convergence. His overtly class analysis demonstrates a change in the concept of privacy wrought by the practice of productive labor as everyday leisure activity.

From popular culture to scientific research to border regulation, governance, entertainment, production and consumption…almost every identifiable facet of human and posthuman life has been affected by a "network" paradigm.

From fringe groups to medical collectives, from Health 2.0 to Kiva.org, from sports, to travel, to television, everyday experiences have shaped what we understand as life in the network. Therefore, chapters in this volume work to bridge critical theory and the practice of everyday life, examining how politics can be conducted, bodies cared for, and identities managed within life in the network. In sum, the chapters herein make a proposition that any critical examination of the network will benefit from paying attention to examples drawn from everyday experience. By placing emphasis on a critical textual analysis of particular case studies, we are beginning the work of operationalizing network as a variable and preparing it for future use. As a cohesive whole, this collection asks the question: How does one conceptually use "network" to analyze evolutions in critical cultural discourse and the everyday practices that it addresses?

NOTES

1. These particular post 9/11 texts altered the traditional 20th century technological anxiety narrative from focusing on a singular, centralized and identifiable digital protagonist represented by the mainframe computer, to a concept of ambiguous, distributed and multiplicious protagonism represented by network.
2. There are a few different uses of the term 'network' in play in this book. There is 'network' as a condition of the everyday life. There is 'the network', which is the actual agglomeration of nodes/points that constitute the global network array. And then there is 'a network', which a singularized set of nodes/points that stands outside of but in relation to other networks.
3. See Kien (2009a).
4. The virus causes the infected to run violently amok and further spread the virus through direct contact such as biting, bleeding and salivating.
5. For a summary, see Kien (2009b).

REFERENCES

28 Days Later. (2003). Danny Boyle (director). Fox Searchlight Pictures.

28 Weeks Later. (2007). Juan Carlos Fresnadillo (director). 20th Century Fox.

Baudrillard, J. (2002). *The Spirit of Terrorism*. New York: Verso.

Baudrillard, J. (1988). *The Ecstacy of Communication*. Bernard & Caroline Schutze (trans). Sylvére Lotringer (ed). New York: Autonomedia.

Castells, M. (2000). *The Rise of the Network Society, 2nd edition*. Malden: Blackwell.

———. (2001). *The Internet Galaxy: Reflections on the Internet, Business, and Society*. New York: Oxford University Press.

———. (co-editor). (2006). *The Network Society: From Knowledge to Policy*. Washington, DC, Center for Transatlantic Relations.

Craven, P., and B. Wellman. 1973. "The Network City." *Sociological Inquiry* 43:57–88.

Galloway, A. R. and E. Thacker. (2007). *The Exploit: A Theory of Networks*. Minneapolis:University of Minnesota Press.

Gere, C. (2002). *Digital Culture*. London: Reaktion Books Ltd.

Grewal, D. (2008). *Network Power: The Social Dynamics of Globalization*. New Haven: Yale University Press.

Hardt, M. and A. Negri. (2000). *Empire*. Cambridge, MA: Harvard University Press.

Kien, Grant. (2009a). *Global Technography: Ethnography in the Mobile Field*. New York: Peter Lang Publishing.

———. (2009b). "Hybrid Networks, Relational Materiality". *Material Culture and Technology in Everyday Life*. Phillip Vannini (ed). New York: Peter Lang Publishing.

Latour, Bruno. (1988). *The Pasteurization of France*. Alan Sheridan and John Law (trans.). Cambridge, MA: Harvard University Press.

Law, J. (1992). "Notes on the Theory of the Actor-Network: Ordering, Strategy, and Heterogeneity." *Systems Practice*, 5:4, pp. 379–393.

Terranova, T. (2004). *Network Culture: Politics for the Information Age*. London: Pluto.

van Dijk, J. (1999). *The Network Society: Social Aspects of New Media*. Thousand Oaks: Sage Publications.

Virilio, P. (2002). *Ground Zero*. Chris Turner (trans). New York: Verso.

———. (2000). *The Information Bomb*. Chris Turner (trans). New York: Verso.

Wellman, B. (2001). "Physical Place and Cyber Place," *International Journal of Urban and Regional Research*.

———. (1979). "The Community Question: The Intimate Networks of East Yorkers." *American Journal of Sociology*, *84* (March): pp. 1201–31.

———. (1988). "Structural Analysis: From Method and Metaphor to Theory and Substance." *Social Structures: A Network Approach*. Barry Wellman and S.D. Berkowitz (eds). Cambridge: Cambridge University Press, pp. 19–61.

CHAPTER ONE

When Collective Intelligence Agencies Collide

Public vs. Popular Intelligence and Networked Suspicions

JACK BRATICH

When John Arquilla and David Ronfeldt (2001), in their oft-cited passage, declared that it takes a network to fight a network, they made a link between decentralization (leaderless, antihierarchical tactics) and broad identities (in this case, ideologies). They did so within the framework of warfare (netwar). I argue in this chapter that war is not a special case of networks but a context for understanding their existence. Networks have a lineage in warfare, specifically within counterinsurgency studies and information warfare (Terranova 2004). What are new network enemies? How do they get detected? What war tactics are still in effect and which new ones get developed?

This chapter begins to examine the conditions under which networks can act or be prevented from acting. It focuses on one element of networks, what has come to be called collective intelligence. Using the 9/11 Truth Movement (9/11 TM) as a case study, I examine how a particular networked collective, organized around research and skepticism, is targeted by State and non-State actors. The 9/11 Truth Movement enables us to think about the ways in which networks are embedded in institutional histories. In addition, this case allows us to see one historical example of the politics of everyday life, namely the *networking of suspicion*. In times of terror/war, how are people mobilized to be suspicious subjects? How are peer-to-peer networks of suspicion clashing?[1] I propose thinking of this case in terms of Public versus Popular Intelligence, in which traditional political categories (even if mutated) enable the task of distinguishing among networks.

COLLECTIVE INTELLIGENCE

One way to understand the different ways networked groups interact is via the concept Collective Intelligence (CI). First introduced by French social theorist Pierre Levy, CI refers to the recently emerging forms of technologically enhanced decision-making and community-formation. Collective Intelligence refers to "open-minded, cognitive subjects capable of initiative, imagination, and rapid response" involved in "innovation networks." (Levy 1999, p. 1) Technological developments like distributed computing, online interactions across space, and rapid communications have created enhanced capabilities for problem solving in a wide variety of fields (political, engineering, marketing, cultural). In addition, the failure of institutions and the violence resulting from identity-differences (ethnic, religious, national) have forced the need to develop new models of belonging and organization.

Collective Intelligence's new deliberative body works via distributed knowledge and network forms. We can see it at work in such examples as crowdsourcing, P2P file sharing, management theory, feedback, and interactive marketing. CI is more than a tool or political form for Levy. It is related to the etymological roots of intelligence: *inter legere*, meaning joining together, constructing (p. 10). For Levy, intelligence as networked knowledge production will be the foundation for the reconstruction of the social bond.

Even grander, CI represents a "new stage of hominization." (p. xxvi) It is a *species* response to power inequalities and the ruinous division of intelligence. The "We" produced in this new collective is thus ontological, a species-scale collectivism. But rather than rely on network models to understand a new ontological moment (as a new stage of Being or Humankind), we need to address the reverse, as posed by Alex Galloway and Eugene Thacker (2007)—to understand the ontology of the network itself. One key means of doing this, they argue, is via the political and philosophical task of analyzing how networks individuate or are individuated; how their being, identity and persistence occurs. How can we discern the shape of emerging networks?[2]

The following case study takes on this challenge as it pertains to collective intelligence. Ultimately, I want to argue, there is no single CI; it is the *determination* of this CI that is the key task. Among the tangles of networked hybrids we can tell the tale of one Collective Intelligence Agency (the 9/11 Truth Movement) and its responding actors. This clash of suspicions allows us to think of how networks get individuated, especially via traditional political categories such as public and popular. In addition, the immersion of networks (of suspicion) into everyday life reminds us that warfare remains a significant context that shapes individuation.

Galloway and Thacker foreground this war context when they note the different models of antagonism defining global politics: Symmetrical (Cold War blocs), asymmetrical (network insurgency vs. State actors), and a return to symmetrical (network vs. network) (p. 14–15). It is this last form, networks fighting networks, and the possibility for an asymmetrical relation via what they call the Exploit, that is salient for our case study of the 9/11 TM as a Collective Intelligence Agency.

THE 9/11 COLLECTIVE INTELLIGENCE AGENCY

The 9/11 Truth Movement is a loosely organized international aggregate of different researchers and websites, all committed to the further investigation of the attacks of September 11, 2001 on U.S. targets. It has no headquarters, no central leadership or representatives. While there are key figures doing different tasks, there is no formal body of rule-makers or organizational board. It describes itself as having an "autonomous, decentralized structure" with no chapters or members. No groups, including those with 911Truth in their titles, can claim to represent the movement. Rather there is a central portal, a nodal website that links to others (http://www.911truth.org/).

But this only explains a minor part of the TM. Spatially, the 9/11 TM operates in a *conspiratology archipelago*, taking place in bookstores (e.g. VoxPop), church halls, streets, websites, community fairs, coffeeshops, and the World Trade Center site. It relies on the decentralized (and often free) distribution of multiple forms of media, such as documentaries (*Loose Change* and *9/11 Mysteries*, among others) and print (books, a journal of 9/11 studies, pamphlets, flyers). Researchers operate via distributed knowledge: different people work on different pieces of the 9/11 puzzle, connecting the dots. Its agents spread signals via bumper stickers, graffiti (in bathrooms, on money), chat rooms, and physical disruptions (notably against Bill Clinton and Bill Maher, in one week). One 9/11 researcher even recognizes these efforts as "viral" in form (Jamieson 2005).[3]

There is no central Truth Movement account of the events of 9/11. Adherents vary on their interpretations of how and why the attacks happened. From the portal, we find three key tasks that comprise the movement as an organizing tool: "Research, awareness, litigation." In its decentralized structure, its commitment to knowledge-based organizing, and its tactical flexibility (e.g. swarming newly published stories that affirm the official account) 9/11 TM functions as Levy's CI.[4]

Moreover, dissensus and conflict are endemic in the 9/11 TM.[5] This is not simply a diversity of perspectives. Positions not only contradict one another (e.g. Plane vs. No Plane hitting the Pentagon), they often accuse each other of being disinformation agents. Rather than thinking of this CI as a community of interest, it is more fruitful to think of it as one of what Geert Lovink and Ned Rossiter (2005) call "organized networks." According to them, "Networks thrive on diversity and conflict (the *notworking*), not on unity," and thus it behooves us "to reflect on distrust as a productive principle." (italics added) The 9/11 TM is thus not just a loose, decentralized set of informational tactics. It is an emergent collective body, with a density of relationships whose affective glue is suspicion.[6] With its internal conflicts as well as external antagonism (and the blurring between them), the 9/11 TM operates as an organized network.

This Collective Intelligence Agency does not emerge in a vacuum, or more accurately, it emerges due to a number of vacuums. The 9/11 TM can be said to spread among the ruins of two institutional bodies, both having to do with investigation. First, there is the National Commission on Terrorist Attacks Upon the United States (a.k.a. the 9/11 Commission), whose report was published in July 2004. Much effort has been spent poking holes in the 9/11 Commission Report, concentrating and accelerating the decades of similar critique of another major government report, the Warren Commission. In other words, the charge against the Commission Report is not just about one failed investigation, but the failure of government-appointed investigative bodies to perform above suspicion and free of corruption.

The second institutional body is broader. The TM consistently notes the failure of mainstream or professional journalism to do anything but repeat the official line about 9/11. Again, this is not an isolated incident but a spectacular example of the ongoing erosion of investigative journalism (as an intelligence-gathering practice) and of the routine interlocking patterns of press and government. The resources and institutional affordances for investigation by professional journalism have been degraded, and thus cannot be trusted to have an independent understanding of major events. The 9/11 TM challenges the 9/11 Commission's investigation as well as journalism's investigatory powers, noting that to have truly democratic decision making requires proper investigative bodies and intelligence gathering mechanisms. Among the remains of these professional mechanisms arises the 9/11 TM, which as an organized network is a "hybrid formation: part tactical media, part institutional formation." (Lovink and Rossiter 2005)

We could call this hybrid formation an *amateur investigation movement*. Its collective intelligence, while often relying on experts and institutionally accredited researchers, does not depend on an institutionalized authorization for its practices. For some, this automatically disqualifies any claims emerging from this movement (How can you trust it? Who is regulating the veracity?). But rather than equate authority with professional authorization, we can link 9/11 TM to a wider array of amateur knowledge-production. Indymedia, blogging, citizen journalism, and social movement media can be understood as parallel amateur agencies in this milieu, situated in a broader history of street-based knowledge.[7]

These amateur knowledge networks contribute to what Henry Jenkins (2006), analyzing popular culture via Levy (among others), calls a *popular epistemology*. These are ways of knowing that bypass officially sanctioned means of apprehension in order to ground interpretations in everyday life. More specifically, the TM creates a type of what John Fiske calls "counterknowledge": it connects the dots, fills in gaps, and creates a contextual analysis that explains events and anomalies in opposition to officially sealed and hegemonic accounts (Fiske, 1994). Thus, these *popular investigations*, as we could call them, are fueled by a suspicion (of the official account and the alleged independent bodies that investigate it). Popular suspicion, at least in part, individuates this collective intelligence agency, giving it an object (official accounts) and adversaries (decaying investigative bodies).

At the same time, this popular epistemology becomes the target of another kind of suspicion, one that determines the 9/11TM to be fueled by conspiracy theory, to be paranoid, and even to be a threat. Unofficial accounts of 9/11 have been pilloried by dominant news outlets as well as more radical Left and Right media. Soon after the attacks, George W. Bush announced, "Let us not tolerate outrageous conspiracy theories." Around the fifth anniversary of the attacks, numerous Left commentators ritually piled on their criticisms (Bratich 2008, pp. 140–148).

More importantly, in the mid 2000s, the US State Department released its *National Strategy for Combating Terrorism*. This strategy document, among other things, seeks the root causes of terrorism. Among the list offered is the claim that "subcultures of conspiracy and misinformation" are one of the sources that terrorism "springs from." While not naming the 9/11 TM explicitly, it is no stretch to place it within one of these "subcultures." This is especially evident in light of another governmental initiative, also located in the State Department. The Bureau of International Information Programs launched a website in July 2005 called *How to Identify Misinformation*. Designed for journalists and everyday citizens, the manual's first section gives the following tip for detection: "Does the story fit the pattern of a conspiracy theory?" and provides a number of elements to look for. In January 2006, the top misinformation story that needed attention was September 11 conspiracy theories. Later in the year, the State Department updated the project with a list of the Top 9/11 conspiracy theories (2006).

Putting these strategic documents together, we can say that the State Department is recruiting a public (and its journalistic proxies) into the Terror-War as information monitors. One of the public's key tasks is to identify accounts as conspiracy theories. This detection is crucial not because the CTs are simply false, but because they are a danger (fueling a subculture which is a source of terrorism).

The monitoring of citizen thought is a component of what has been called the New Normal. Mark Andrejevic (2006) analyzes this context, in which Terror War depends on a governmentalized conduct of conduct in the spaces of everyday life (e.g. train platforms, buildings).[8] Specifically, citizens are recruited via a "lateral surveillance" which includes obvious campaigns like "If you See Something, Say Something" in addition to more subtle (but pervasive) cultural forms like reality television programs, online social network interactions, and infidelity detection technologies (Andrejevic 2006).

Enrolling citizens in this Terror-War is but the most recent example in a long history of recruiting citizens as spies and dissent-monitors (Redden 2000). Having citizens keep an eye out for conspiracy theories as dangerous thought-forms (leading to criminal action) was discussed prominently soon after the 1995 Oklahoma City bombing. During this time (an acute phase of what I call "conspiracy panics") individuals were encouraged to *detect* and *identify* political paranoia, mobilizing citizenry for antiextremism and antipopulism campaigns.

As a crucial component of snitch culture, the 1990s monitoring of hate and extremism included an incitement of popular suspicion directed at fellow citizens in a kind of democratization of dissent-management. Monitoring dangerous forms of thought was promoted as a citizen's duty. This decentralized, p2p suspicion trained subjects to be suspicious of suspicion, to turn critical impulses against criticism, in the name of a moderate rational order. The popular was mobilized against the popular in the name of a public.

Citizens turn their suspicion on each other to root out possible inclinations towards conspiracy theories. The citizen subject is a peer detector, one that acts as a conductor and director in a network of suspicion. Moreover, while this network had been emerging in the mid to late 1990s primarily in what can be considered "civil society," the post-9/11 arrangement sees an explicit coordination and anchoring of this network by an executive State (e.g. the State Department projects).

In sum, we are faced with two peer-to-peer *networks of suspicion*. How are they to be distinguished? From one angle, we have the 9/11 TM and other such narratives skeptical of the official account. They direct their suspicion at State action (domestic and foreign), White House statements, mainstream journalism's accounts, and sanctioned investigative bodies. This skepticism addresses not only the events of 9/11 and their wake, but also a broader context of US history, foreign policy, domestic operations, and social control.

From another angle, we see a network mobilized and directed against the first one. A peer-to-peer network in the name of a public, against the network of suspicion deemed to be an excessive suspicion, a conspiracy theory, an exaggerated pretender, para-noid.[9] The direction of suspicion in this network depends on a subject (the monitor) identifying with State objectives. This CIA (Collective Intelligence Agency) is thus woven into its acronym's institutional predecessor.[10]

Are these separate networks? Not at the outset. We can instead discuss them as asymmetrical and antagonistic. Each is anchored in suspicion. But this common point is also the point of rupture and divergence. They do not occupy the same space nor do they face each other. Rather, they accumulate their mechanisms and gather strength or become weakened vis-a-vis the other. How do they gather their forces?

While these might not be separate networks, each with its own terrain or ground of being, we can begin to distinguish them by focusing on how networks get mobilized around *recruitment*, *identification*, and *objectives*. In other words, contextualizing networks within warfare brings forth not only the question of tactics but of adversaries—the Us and Them. Even the oft-cited Rand quote about network-centric warfare comes from a context in which enemies are either presumed (in the Cold War case: Soviets and their distributed Marxist mole agents) or need to be identified (in the post Cold War context: terrorism). No wonder the State Department had to compile and publicize a list of terrorist organizations to ensure a proper focus for understanding network warfare.

In addition, the terrorism designation can be applied to domestic actors, as demonstrated in the case of the terrorism-enhancement sentencing of US environmental activists. As a number of analysts note, making terrorism a target allows for the creation of "unspecified enemies," a position whose content can be filled in by "whatever enemy" those with discursive and material power so choose (Deleuze and Guattari 1987; Debord 1998; Massumi 1993; DerDerian 2001). How subjects identify with these forces, these creations of Us and Them, is key to discerning networks. Networks still form within directions, positions and objectives, and thus we can begin differentiating network hybrids around such characteristics.

PUBLIC VS. POPULAR INTELLIGENCE AGENCIES

As an initial attempt at understanding these matters of distinction, I want to propose one conventional binary to make sense of the tangle of Collective Intelligence Agencies, namely *Public* and *Popular*. I have already referred to the

9/11 TM as a type of popular epistemology. I have also noted its characteristics as a popular investigative agency, via the amateur practices swarming through the ruins of professional bodies. Popular here does not simply mean a pre-existent "people," nor something that is necessarily widespread or common. Instead, it refers to the way Michel De Certeau (1984) discusses popular culture as existing in interstices in ways that subvert or avoid dominant culture (akin to guerrilla warfare). It also invokes Michel Foucault's (1980) discussion of popular knowledges as being "subjugated knowledges": below the threshold of scientific belief and outside of the dominant regime of truth.

Finally, I am using popular here to refer to how cultural studies has discussed it in reference to populism.[11] The popular does not simply presume a people or a common knowledge, but seeks to shape and construct one. The "people" is up for grabs, and is a project to be articulated and enacted. Here again we can link this project of constructing a people to the history of insurgency and guerrilla warfare, as a "people's war." Popular suspicion can thus be defined as skepticism against one articulation of Us/Them (e.g. "the Public vs. extremism"). Popular Intelligence places the question of identification with positions and objectives at the center of its project.

However, this popular is not the same as the modern category of "the people," forged as it was in the crucible of the nation-state. Rather, this networked populism is closer to the political formation that recent autonomist social theory has called the "multitude." (Hardt and Negri, 2000, 2004; Virno 2004; Dyer-Witheford 1999; Neilson and Rossiter 2005) In order to devote more attention to Public Intelligence, I will only note three elements that link the popular to the multitude here: 1) An ethos of recognizing and negotiating internal variation (e.g. the disunity and asymmetry that defines organized networks, or the "respect for singularities"); 2) the beginnings of institutionalization (or at least a sedimentation of practices comprising a nascent decision-making body); and 3) the constitution of enmity (or how a commons addresses its own "constitutive outside") (Rossiter 2006).

By the Public I am referring to two of its historical meanings: first, as the central category of political subjectivity (the *publius* of classic political thought) especially in modern republics. This public is typically invoked by conspiracy panic discourses in order to guide the thought and conduct of citizens. Organized around antiextremism, this public refers to the moderate and rational forms of knowledge and conduct comprising citizenship. It is a dominant "Us" which invites subjective identification in order to detect "Them."

The second version of the Public is more germane here. It is drawn from the early to mid 20th century emergence of Public Relations and Public Opinion. A number of researchers note that these constructions of the public emerged as a result of the successful propaganda efforts during World War I

(Ewen 1996; Stauber and Rampton 1995). A key figure here is Edward Bernays, regarded as the father of PR, and an important member of the Committee on Public Information (aka the Creel Committee) whose tasks were to mobilize the recalcitrant US population into war. Bernays took some of the propaganda techniques (creating images of vicious enemies, using psychology to induce states of guilt and anxiety in the target population) and applied them to the burgeoning field of PR. Similarly, these and other practices were formally studied and refined in newly created institutions for the study of persuasion, public opinion, and communications (Simpson 1994).

Managing the growing power of masses was paramount, which meant devising means of recruiting, directing, and harnessing them (Ewen 1996). The "public" became a keyword for this managed population. The public was to be both domesticated and persuaded (to align with a position)—an identification and a recruitment around an Us/Them—in its early case around State warfare objectives. These central imperatives were implemented in covert decentralized ways as well, such as the Four-minute Men, who would make seemingly impromptu patriotic speeches at town meetings and civic association gatherings.

At the same time that a public was being mobilized through persuasion, other subjects were being immobilized through *dissuasion* (Virilio 2000). The Palmer Raids, the Espionage Act, and the Sedition Act all worked to ensure that only particular opinions and actions would be operational in the name of the public. Throughout the 20th century similar practices comprised a history of domestic enemy production, from immigrant radicals in the 1920s and 30s to the Red Scare over fifth column communists in the 1950s to domestic extremists in the 1960s and 1990s (Rogin 1987). In other words, constructions of a public, a "We" at the center of US political life, have persistently depended on creating an enemy relation with some domestic dissenters. This public arrives on the scene only via the management and demonization of dissent (domestic threats who end up functioning as a Them among Us). This is how "terrorism" does not simply belong to an easily marked foreign Other but can appear anywhere (the unspecified enemy). The Post 9/11 Terror/War US initially directed its enmity at the Arab or Muslim Other (and even this crossed borders). However, very quickly the enemy became quasi-open-ended (whose content could range from Marxist guerrillas in Philippines to eco-activists damaging property in the Northwest US).

Producing a *public* thus requires dissent monitoring; mobilization thus entails immobilization. And it is this dissuasion of action that is key to understanding the different networks of suspicion in a contemporary warfare context. One way to comprehend this significance is to turn our attention briefly to another recent example of a CIA as a peer-to-peer suspicion network, Open

Source Intelligence.

About a month after 9/11/01, former clandestine intelligence case officer Robert Steele made an appeal for a "new craft of intelligence." He called upon Congress and to the public to bolster what he named "Open Source Intelligence." OSI would entail analysis of publicly available documents (news reports, commercial imagery, TV stories, Internet) and produce "a nation in which each citizen is both a collector and consumer of intelligence." (Steele 2006) Steele suggested the creation of a national Open Source Agency. He got his wish when, following the recommendations of the 9/11 Commission, in late 2005 the Director of National Intelligence announced the creation of the Open Source Center. The Center was established to collect information available from "the Internet, databases, press, radio, television, video, geospatial data, photos and commercial imagery." (Office 2005) It would involve, among other things, amateur spotters using Google Earth to monitor conduct.

A couple of years later, Steele called for an end to the OSC. Steele then went on to edit and publish a collection bearing the same name as Pierre Levy's groundbreaking book: *Collective Intelligence*. Interestingly enough for our purposes here, on the first page of the preface, Steele announces a proposed substitute name for OSI: *Public Intelligence*.

The name change significantly signals an identification with an "Us," but does not erase the previous tactics and objectives. It is still a means of recruiting citizens to do monitorial work, with the primary goal of producing *actionable intelligence*. At one level, it means turning the glut of publicly available information into action. This is central to classic information warfare and intelligence warfare. Translating information into action under the sign of the Public, however, has another dimension. Given our earlier discussion of the history of dissent-management, we need to focus on the dissuasion that accompanies persuasion, the immobilization that shadows mobilization.

Tiziana Terranova brings this to our attention regarding networks. In a Rand anthology edited by Arquilla and Ronfeldt (of "it takes a network to fight a network" fame), John Rothrock defines infowar as "the degradation of adversaries' capacity for understanding their own circumstances, but also *the capacity to neutralize any effective use of whatever correct understandings they might achieve*." (Rothrock 1997, quoted in Terranova 2007, p. 132; italics added) For Terranova, this component of information warfare is indispensible for comprehending what she calls a "futurepublic" (2007). A collective intelligence is thus inextricably linked to its *agency*, here defined as the ability to act on its information. CIA is more than a play on the acronym for a State Department institution devoted to the craft of intelligence. It points to the very capacity to render information actionable, to do something with knowledge once attained.

Here we return to the importance of distinguishing and individuating networks of suspicion. One CIA seeks to render the other inactive, to neutralize its agency. Under the name Public Intelligence, peer to peer suspicion creates an Us via recruiting a population to *identify* with state security imperatives (monitor behavior and speech of others). It antagonistically substitutes one suspicion for another.[12] In doing so, it directs its skepticism laterally, ensuring that no extremist forms of skepticism are able to act on their understanding. The 9/11 TM, no matter how it might achieve some compelling sense of understanding, will not get any traction, as its ability to act on such information (in the form of organizing, mobilizing) is itself the target of the monitoring. Preventing people from identifying with "conspiracy theories" does not primarily diminish the *understanding* that might be achieved by those in the 9/11 TM (that degradation might come from other covert operations like infiltration and disinformation), Rather, this dissuasion decomposes the ability to spread that understanding, to act as a social movement, to mobilize.

Targeting one network of suspicion with another around matters of proper dissent and rational political thought disrupts, preventively, the constitution of a *popular*. The popular here can be the intransigent force refusing to be recruited into State identification, especially war projects. Or it can be the project of constituting new norms and practices (e.g. of investigation) amidst the ruins of established institutions. Ultimately what is being prevented is the reconstitution of an Us/Them, a network experiment in modes of individuation. Pierre Levy (1999) notes this pre-emptive tendency: "power must continuously thwart the emergence of a collective intelligence that would enable the community to forsake such a power." (p. 82) In this way, Public Intelligence (as a CIA) is an interference of transmission, a pre-emption of agency, and a disruption of composition; in sum, an antagonistic relationship to its outside.

Agency, the capacity to render information actionable, is thus the very target of one CIA by the other. One mode of individuation (the public) depends on the neutralization of another (the popular) in order to achieve its own capacity to act. Rather than think of one network spatially supplanting or eliminating another, it is more fruitful to think of it temporally: the constitution of Public Intelligence is predicated on a pre-emptive management of dissent, a thwarting of emergence, a divisiveness that prevents some virtuals from becoming actualized.[13] It is not a type of static warfare model whereby a subject is completely removed from the battlefield. Rather, their capacity for action, for increasing their effectiveness by joining powers, is diminished in a dynamic manner. Dissuasion can come via criminalization, but doesn't need to take such codified and rigid forms. Popular intelligence becomes a "subjugated knowledge" by official sources, by journalists and pundits, by popular cultural depictions, and other cultural agents. Rather than making Popular Intelligence disappear, the objective is similar to the one elaborated by Rothrock regarding information warfare: defusing and depriving agents of sources that would increase their strength, thereby rendering them incapable of action.

CONCLUSION

How is it, then, that networks individuate? Here I have proposed a way of understanding network hybrids and entanglements through one conceptual distinction (public and popular as antagonism). This differentiation arises from a particular context for understanding collective intelligence, namely warfare. This history is not just a genealogy of campaigns, weapons, and tactics; it is primarily a series of alliances and enmities, a construction of an Us/Them. How do these subjects persist or mutate under a reconstitution of the "collective"?

Popular intelligence directs skepticism against one organization of Us/Them, and places the question of identification with objectives and positions at the center of its network of suspicion. Public Intelligence, or governmentalized networks of P2P suspicion, increases in proportion to the immobilization of popular intelligence. Distinguishing public from popular is thus not just an analytic formal matter, but also a political one, involving issues of sovereignty, agency, and the capacity to expand one's own powers. Networks can be distinguished around Us/Them and other such bulky political categories, even while in practice there is significant overlap among actors and nodes.

The two CIAs here form overlapping while antagonistic dual networks of skepticism. But is this a "Clash of CIAs" or one network? As Galloway and Thacker note, networks are internally heterogeneous, containing within them "antagonistic clusterings, divergent subtopologies, rogue nodes." (p. 34) These even include "incompatible political structures." (p. 34) On one level, the 9/11 Truth Movement itself contains these incompatibilities in the form of differing ideologies, styles of organizing, and perhaps even motivations (e.g. disinformation).[14] And when the 9/11 TM gets taken as a target of State-coordinated networks of suspicion, another layer of heterogeneity erupts. How does Public Intelligence encounter Popular Investigation? What is the network that contains and fuses these two? What name could we give to this emergence? Is this a new CIA?

While located in a longer historical context, both CIAs take shape through the rupture of 9/11, such that neither is completely separate, nor is one an original and the other a mimic. But they emerge from different conditions, each defined as a failure. On the one hand, the "failure" of intelligence (as articulated by the 9/11 commission) leads to calls for better intelligence operations, such as OSI. On the other, it is the failure of investigative bodies themselves (journalism and the 9/11 Commission itself) that spurs the growth of the 9/11 TM.

In each case, we see an attempt to recompose sovereignty and subjectivity in newly organized societies of security.

Both the 9/11 TM and the Public Intelligence project harness the powers of collective intelligence, sometimes even using the phrase. The networks entangle and mutate in their tactics, their recruitment efforts, perhaps even their agents. A telling final anecdote might illuminate this overlap. A tiny bookstore/cafe in NYC's East Village used some of its few shelves to display prominently Robert Steele's *Collective Intelligence* collection when it was first published. What made this scene remarkable was that the store was VoxPop, a key informational outlet for 9/11 Truth Movement research![15] Here was *Collective Intelligence*, penned by a former intelligence agent, in the heart of one of the key sites in the 9/11 CIA archipelago. Which CIA is involved here?

NOTES

1. For a sustained discussion of peer-to-peer, lateral surveillance, see Andrejevic 2007.
2. While this chapter addresses Levy's species abstractness by situating CI in politics, Galloway and Thacker offer a more ontological approach by highlighting the nonhuman, even misanthropic aspects of networks. Rather than think of CI as another stage of hominization, we might be witnessing another mutation altogether, one that belongs to something more elemental.
3. These viral or guerrilla tactics can be said to be borrowed from marketers, who themselves appropriated the techniques from activists and warfare strategists.
4. As I note elsewhere, this decentralized component mixes with more traditional forms of organizing, complete with local leaders, hierarchically run meetings, and mass mobilizations (Bratich 2008).
5. For an analysis of some of these internal divergences and their impact on attempts to form a conspiracy community, see Fenster 2008. Moreover, the work examines how the 9/11 Commission mobilized rhetoric in advance of these conspiratological criticisms to ward off comparisons to the Warren Commission.
6. One could go further and say that it carries the burden of suspicion, one that has been displaced and expelled by the hegemony of a particular 9/11 account. A political analysis of this collective affectivity (key to understanding any collective intelligence) might even understand this burden as an embodiment to the point of toxicity. In other words, popular suspicion, left without outlets for expression, turns on itself in depleting and poisonous manners. We might even see this as an affective exploit—the inhuman element of a networked suspicion that carries with it the demise of the agency founded on it.
7. See Soderlund 2002. At times, examples of this street knowledge are quite literal, such as the sale of conspiracy research literature by street vendors in Harlem and near the World Trade Center. At other times the metaphor persists, as when political paranoia is associated with "Arab Street."
8. See also Bratich (2009); Gates (2006); Hay (2006); Packer (2006).
9. In *Conspiracy Panics*, I argue that conspiracy research is regularly defined as a hypersuspicion and a simulation of normal suspicion. It is thus treated as a danger because it is prox-

imate to the norm while seeming far away (the "para" in "paranoia").

10. It is important to note here that the State Department formulations were only in existence during the Bush administration. As of this writing, it is too early to tell what the Obama model of information warfare and dissent-management will look like regarding networks. However, early stirrings indicate a return to one of the Clinton-era targets: cyber-attacks.

11. Here Mark Fenster's (2008) work on conspiracy theories as engaged in a kind of populist politics is key.

12. While there is no space here to elaborate, it is important to note here that this substitution is a mimicry and mutation, a replication that has roots in virology as well as warfare.

13. For more on this preventive model of dissent management, see Elmer and Opel (2008).

14. In the very spirit of what Levy calls an "apprenticeship in civility," the 9/11 TM has attempted to mitigate some of these rancorous differences via their "Commitment to More Civil and Effective Collaboration in the 9/11 Truth Movement."

15. This store was a smaller satellite of the main Brooklyn shop. A few months later the Manhattan outlet closed down.

REFERENCES

911truth.org. "Commitment to More Civil and Effective Collaboration in the 9/11 Truth Movement" http://www.911truth.org/page.php?page=civil_commitment

Andrejevic, Mark. (2006). "Interactive (In)Security," *Cultural Studies*, vol. 20,4–5, pp. 441–458.

———. (2007). *iSpy: Surveillance and Power in the Interactive Era*. Lawrence: University Press of Kansas.

Arquilla, John, and Ronfeldt, David. (2001). *Networks and Netwars*. Santa Monica, CA: Rand.

Bratich, Jack. (2008). *Conspiracy Panics: Political Rationality and Popular Culture*. Albany, NY: State University of New York Press.

———. (2009). "Spies Like Us: Secret Agency and Popular Occulture" In J. Packer (Ed.), *Secret Agents: Popular Icons beyond James Bond* (pp. 133–162). New York: Peter Lang.

De Certeau, Michel. (1984). *The Practice of Everyday Life*, trans. by S. Rendall. Berkeley: University of California Press.

Debord, Guy. (1998). *Comments on the Society of the Spectacle*. London: Verso.

Deleuze, Gilles. & Guattari, Felix. (1987). *A Thousand Plateaus*, trans. by B. Massumi. Minneapolis: University of Minnesota Press.

DerDerian, James. (2001). *Virtuous War: Mapping the Military-Industrial-Media-Entertainment Network*. Boulder, CO: Westview Press.

Dyer-Witheford, Nicholas. (1999). *Cyber-Marx*. Urbana: University of Illinois Press.

Elmer, Greg and Opel, Andy. (2008). *Preempting Dissent: The Politics of an Inevitable Future*. Winnepeg, MB: Arbeiter Ring.

Ewen, Stuart. (1996). *PR! A Social History of Spin*. New York: Basic Books.

Fenster, Mark. (2008). *Conspiracy Theories: Secrecy and Power in America*. Minneapolis: University of Minnesota Press.

Fiske, John. (1994). *Media Matters: Everyday Culture and Political Change*. Minneapolis: University of Minnesota Press.

Foucault, Michel. 1980. "Two Lectures," In C. Gordon (Ed.), *Michel Foucault: Power/Knowledge* (pp. 78–108). New York: Pantheon.

Galloway, Alexander R. and Thacker, Eugene. (2007). *The Exploit. A Theory of Networks*.

Minneapolis: University of Minnesota Press.

Gates, Kelly. (2006). "Identifying the 9/11 'Faces Of Terror': The Promise and Problem of Facial Recognition Technology," *Cultural Studies*, vol. 20, 4–5, pp. 417–440;

Hardt, Michael and Negri, Antonio. (2000). *Empire*. Cambridge, MA: Harvard University Press.

———. (2004). *Multitude*. Cambridge, MA: Harvard University Press.

Hay, James. (2006). "Designing Homes," *Cultural Studies*, vol. 20,4–5, pp. 349–377.

Jamieson, Les. (2005, September 7). What Action Looks Like: Going viral with 9/11 truth. Retrieved January 4, 2006, from http://www.ny911truth.org/articles/what_action_looks_like.htm

Jenkins, Henry. (2006). *Convergence Culture: Where Old and New Media Collide*. New York: New York University Press.

Levy, Pierre. (1999.) *Collective Intelligence: Mankind's Emerging World in Cyberspace*. New York and London: Plenum Trade.

Lovink, Geert and Rossiter, Ned. (2005). "Dawn of the Organised Networks." *Fibreculture*, vol. 1, 5. http://journal.fibreculture.org/issue5/lovink_rossiter.html

Massumi, Brian. (1993). "Everywhere you want to Be: Introduction to Fear," In B. Massumi (ed.), *The Politics of Everyday Fear*. Minneapolis: University of Minnesota Press.

Neilson, Brett and Rossiter, Ned. (2005). "Multitudes, Creative Organisation and the Precarious Condition of New Media Labour," *Fibreculture*. vol. 1, 5. http://journal.fibreculture.org/issue5/index.html

Office of the Director of National Intelligence. "ODNI Announces Establishment of Open Source Center." Press release, 8 November 2005.

Packer, Jeremy. (2006). "Becoming Bombs: Mobilizing Mobility in the War of Terror," *Cultural Studies*, vol. 20,4–5, pp. 378—399.

Redden, Jim. 2000. *Snitch Culture*. Venice, CA: Feral House.

Rogin, Michael. (1987). *Ronald Reagan: The Movie*. Berkeley: University of California Press.

Rossiter, Ned. (2006). *Organised Networks: Media Theory, Creative Labour, New Institutions*. Rotterdam: NAi Publishers.

Rothrock, John. (1997). "Information Warfare: Time for Some Constructive Skepticism," in John Arquilla and David Ronfeldt (Eds.) *In Athena's Camp: Preparing for Conflict in the Information Age*. Santa Monica, CA: RAND Corporation.

Simpson, Christopher. (1994). *Science of coercion: Communication research and psychological warfare 1945–1960*. New York: Oxford University Press.

Soderlund, Gretchen (2002). "Covering Urban Vice: The *New York Times*, White Slavery, and the Construction of Journalistic Knowledge," *Critical Studies in Mass Communication*, vol. 19,4, pp. 438–60.

Stauber, John. and Rampton, Sheldon. (1995). *Toxic Sludge Is Good for You: Lies, Damn Lies, and the Public Relations Industry*. Monroe, ME: Common Courage Press.

Steele, Robert D. (2006, April 18). "Open Source Intelligence," *Forbes.com*. Retrieved April 19, 2008 from http://www.forbes.com/2006/04/15/open-source-intelligence_cx_rs_06slate_0418steele.html

Terranova, Tiziana. (2004). *Network Culture: Politics for the Information Age*. London: Pluto.

Terranova, Tiziana. (2007). "Futurepublic: On Information Warfare, Bio-Racism and Hegemony as Noopolitics," *Theory, Culture & Society*, vol. 24, 3, pp. 125–145.

US Department of State. 2005. How to Identify Misinformation. http://usinfo.state.gov/media/Archive/2005/Jul/27–595713.html. Retrieved December 15, 2005.

US Department of State. 2006. The Top September 11 Conspiracy Theories. http://usinfo.state.gov/xarchives/display.html?p=pubs- Retrieved November 7, 2006.

US Department of State. National Strategy for Combating Terrorism. http://www.state.gov/

s/ct/rls/wh/71803.htm

Virilio, Paul. (2000). *Strategy of Deception*. London: Verso.

Virno, Paolo. (2004). *A Grammar of the Multitude*. Los Angeles, CA: Semiotext(e).

CHAPTER TWO

Intellectual Inquiry in the Age of the Efficient Network

Not Unpacking the Infinite Library with Walter Benjamin

JAMES SALVO

I

All my memories of early childhood are very hazy. Then, when I was about nine, they discovered that I badly needed glasses. Apparently, the very young Walter Benjamin badly needed glasses also. He got those and a specially prescribed desk. You can read about it in a vignette in the 1932–34 version of *Berlin Childhood around 1900*, one of my favorite books. Though it isn't this one, there's a vignette that I've been reading over and over in what's considered to be the final version. It's titled "Boys' Books."

"My favorites," Benjamin writes, "came from the school library. They were distributed in the lower classes. The teacher would call my name, and the book then made its way from bench to bench; one boy passed it on to another, or else it traveled over the heads until it came to rest with me, the student who had raised his hand. Its pages bore traces of the fingers that had turned them." (Benjamin, 2006, p. 58)

There's a fitting description here of information transmission, one that we might apply to the notion of *network*. It is only after the boy is *addressed* by the teacher that the book makes its way to him via the mechanism of tiny hands. I'll have more to say on this shortly. What I find more interesting, though, is the last sentence I quoted: "Its pages bore the traces of the fingers that had turned them." We might assume that these traces are not actual fingerprints, ones made, let's say, from the residual stickiness of *ein Berliner*, the pastry. Because this is around 1900, we might guess that Benjamin is referring to page yellowing. Acid-free paper, of course, wasn't commonly used for book printing until the time around *my* childhood, the 1970s, and one reason why the older paper yellows is that it reacts to the oils on your fingers. Fingerprints tell you who touched what, but yellowing only tells you that the book was touched, that its pages were turned. I believe this might tell us something about what happens to intellectual inquiry when the network becomes fully efficient.

The passage quoted above, as I mentioned, is from the final version of *Berlin Childhood*. It's presumably a revision or a replacement of the vignette from the 1932–34 version, "School Library." At first glance, Benjamin's tone seems different:

> It was during recess that the books were collected and then redistributed to applicants. I was not always nimble enough on this occasion. Often I would look on as coveted volumes fell into the hands of those who could not possibly appreciate them (Benjamin, 2006, p. 143).

What we have here is the stealing of enjoyment. It is young Walter who has the wherewithal to appreciate these books, yet they fall into the hands of his philistine peers. In the final version, there isn't a blatant dismissal of his peers' abilities, yet we might unveil a similarity if we compare the two closely. Notice that in the final version the pages are "turned." This doesn't mean, necessarily, that they were read. If we take both passages together, then there is some worry over the stealing of enjoyment, but in the final version, it is by an indeterminate other, by an unaddressable other who turns pages. And it is precisely this idea that I'd like to explore with regard to what might happen to intellectual inquiry in the presence of the post-global network: What is being stolen from the practice of intellectual inquiry with the coming of the infinite library of the network?

II

Technologies have their greatest impact when they become mundane, when our day-to-day tasks become so dependent upon them that, for the most part, we forget we're using them. I use the term *mundane* here deliberately. In other words, technologies have their greatest impact not only when they are *everyday*, but when they become of the world. The term *post-global* that's used in the book title under which this chapter is collected can mean at least two things. It can mean either that we find ourselves in a condition in which the global is no more, or it can mean that we find ourselves in a time that's after the event of globalization. I have no intention of disambiguating the usage of this term for my fellow contributors, but I will disambiguate the term for myself. I prefer the latter sense. We find ourselves in the post-global because globalization is no longer merely emerging, but has entered a stage of perfected incompletion, something we shall explain below. We might think of globalization as that which is world forming, and we find ourselves in a world that will have been formed in a particular way, a world that is, as Jean-Luc Nancy puts it in *The Creation of the World or Globalization,* "the common place of a totality of places." (Nancy, 2007, p. 42) We find ourselves in a world that will have been made by an efficient network. But what is this network, and why is it efficient? Some defining is in order.

I take the network to be infinite. I'm not claiming, though, that what's characteristic of the network is that it's ever expansive or that its principle characteristic is that it's additive. This is to fall into the trap of a spurious infinity. It's the misconception of something being infinite because we can forever add something to it. What makes something infinite, rather, is unbinding. Following what Hegel writes in *The Science of Logic,* something infinite is something in which the finite, or that which binds, has vanished (Hegel, 1969, p. 138). Because I take the network to be infinite, I am claiming that the network is a structure by which things have become unbound.

When something is bound, it's bound by something. It encounters a limit because of that something. Further, if a bound something isn't bound *by* another thing, then it's bound *to* its location. A thing cannot be bound by itself in the capacity of another thing, for that would be to assert that the thing is itself and something other than itself. In other words, a thing can only be said to be bound by itself if we think of binding in terms of encountering a limit. If something is to encounter the limit of nothing other than itself, then it can only encounter that limit because its existence is contained by a particular there. Something bound by nothing other than itself encounters the limit of its there. That something is bound to its location. The structure of the network unbinds these things that are bound to their location. It unbinds them because the network is primarily an index. The network, like in the example from Benjamin's "Boys' Books," facilitates transmission because it makes use of the *address*. Locations are pointed out and named.

We might be accustomed to think of the network as consisting of nodes and connections, but I'm suggesting that nodes and connections aren't what makes the network a network; they are not the essence of the network, for they are not contained within the network, but instead exist outside of it. Though we may have the potential to connect to an almost innumerable number of other nodes, on the level of the everyday, we tend to actualize only a nearly negligible fraction of that potential. I can connect, for instance, to millions of websites or mail a parcel to almost anywhere I can visit on the map, but I never surf the entire net, and I rarely mail anything or travel outside a forty-mile radius of my home.

What's important about the network is the potential it generates, not the actualization of that potential. We may feel as though the world is getting smaller, but it's still gigantic. What we're actually experiencing isn't a shrinking of the world, but the growing accessibility of things located in the world of the network. If one isn't convinced, imagine if we were to surf all websites, mail parcels to all addresses, and go everywhere marked out by a map. Though the network is massive, not all nodes and connections are continuously accessed. This would mean that the size of the world would be in a perpetual flux tied to activity only, not countable locations. The world does not only consist of what nodes are interacting at any one moment through particular connections. What is massive about the network is the collection of what's indexed, what's pointed to, what has an address. If the network has become so massive that it has created the world, then that network is efficient. It is efficient in terms of a *causa efficiens.*

The network creates the world by its gathering together of things to which it gives an address, through indexing. Because the network creates the world through indexing, we can better understand the concept of *network* through a particular part of the network in which the index is the most salient: the infinite library of the network. Though it doesn't yet exist, this networked library is totalized and searchable. It's the size of the world of written texts and is fully indexed. It's a totalized textual archive retroactively created by a concordance—or exhaustive index—of that archive.

III

At this point, not all books are searchable. The technology exists, however, to make this so. Libraries can be digitally scanned, archived, and fully indexed in a search engine. In the efficient network, recorded information itself becomes a totalized and searchable library in which knowledge becomes unbound. The efficient network gives us an infinite library. But what is an infinite library?

One way to think about a library is to think about it as a set of books. To think about a complete library would be to think about a library that contains the set of all books. But what does it mean to think of an infinite library? On the one hand, the answer might seem to be a simple one. An infinite library would be one that contains an infinite amount of books. But here already, the simplicity of the answer starts to break down. We know that this isn't possible, for to *contain* an *infinite amount* would be self-contradictory. As we have already pointed out, what is infinite is by definition without bounds, so it can't be the case that an infinite number of anything can be contained, for to contain something implies that there are boundaries. So within the realm of possibility—both conceptual and material—an infinite library would *seem* to be one that would always allow the addition of one more book. It would be a conception of a library that involves a set that isn't closeable insofar as that set of books cannot contain the last book. In some sense, this is what we have already. The medium—paper, digital, or otherwise—doesn't seem to matter so much.

True, digital libraries seem to lend themselves better to the realization of this type of library than do paper libraries only because paper libraries take up more space, have to be maintained, staffed, *etc.* It isn't impossible, though, to have this type of paper library; it's only difficult to realize. Further, electronic libraries *seem* to be infinite insofar as the number of copies of individual books is not finite. It doesn't contain duplicate copies of books, in fact. Duplicate copies are generated within the point of access—a personal computer, let's say—and there can be as many copies of a particular title as there are points of access. For instance, a library may have ten copies of a certain title, but when those ten are charged, to charge an eleventh book would involve some patience on the part of the prospective borrower. With an electronic library, the only limits to the amount of copies that can be charged are how many points of access exist, but this isn't a limitation of the library itself. The points of access are mere nodes, something we are taking to be different from the infinite library of the network.

Further, it would *seem* that there isn't anything that stops us from writing yet another book. So long as there's some medium upon which and time during which to write, we can always add one more book to the collection of books that have been written already. And though it may someday contain all texts, the infinite library is frozen in incompleteness. It is a perfected or finished incompleteness, something which might seem counterintuitive, but I'll explain this below. This, I think is an important point, the point that with our current conception of the library in general, there's no such thing as a last book. With the infinite library of the network, it isn't a necessary condition that there be a last book, nor is there a promise of a last book. Rather, there's a promise that there will be no last book, but not only because there will always be more books to write and index. There can be no last book, for there are no books in the library of the network.

Let's clarify something, though. The infinite library of the network is created by the index of the network. It's something that's merely gathered together by the index. The network, being an index only, cannot contain the actual books. It's the library—taken merely as a set of books—that would contain the books. However, we aren't talking of just any library. The library of the network, insofar as it results from the network's index, is different. It's different, for that which is indexed becomes transformed. Remember, it isn't only book titles that are indexed by the network, but all text. With the coming of the infinite library of the network, we have a concordance for all public text. For this reason, it's a library that can't be unpacked.

IV

For the bibliophile, Benjamin's essay "Unpacking My Library" is a poignant one. His observations seem true. Property and possession do indeed belong to the tactile sphere, and it isn't difficult to agree that collectors are people with a tactile instinct. True, some of his descriptions may be rare occurrences nowadays. We may no longer be discovering cities, for example, by searching for books in stationary stores or antique shops. Nonetheless, there's something about holding a book in one's hand and owning it. And this might be why one can feel so threatened by the library of the network. Might we feel threatened by the prospect of no longer having actual books, by the prospect of having no libraries to unpack? Yes, it could be this threat to the tactile, to the materiality of books, but this seems to be only half of the story. I think there's something else going on that's parallel. Benjamin's last sentence in the penultimate paragraph is this: "Only in extinction is the collector comprehended." (Benjamin, 2005, p. 492) Similar to this, I think what could possibly be anxiety producing about the library of the network is that the possibility of the extinction of the book is forcing us to comprehend the non-material concept of the book itself. But before we get into this description of the book, we should attempt to address who this collector is, precisely. To understand the collector, we must understand the notion of *collecting*, and the paired notions of *incompleteness* and *completeness*.

> What is decisive in collecting is that the object is detached from all its original functions in order to enter into the closest conceivable relation to things of the same kind. This relation is the diametric opposite of any utility, and falls into the peculiar category of completeness. What is this "Completeness"? It is a grand attempt to overcome the wholly irrational character of the object's mere presence at hand through its integration into a new, expressly devised historical system: the collection (Benjamin, 2002a, pp. 204–5).

It's somewhat ironic that the passage above comes from Benjamin's great, incomplete work, *The Arcades Project*. Still, it gives us a good description of collecting. It's the process of removing an object from the realm of its utility and placing it into relation with things of the same kind, placing it into the category of completeness.

What's more ironic, we might point out, is that to someone like a book collector, the most collectible items are precisely the ones that can't be *collected*, at least in the sense of a gathering together of like objects. To the collector in general, the most collectible are the items that are one of a kind. In classified ads, for instance, we see descriptions like, "Only one of its kind, a rare collectible." So what makes the non-collectible collectible so precious? The attribution of preciousness might be explained as something that's on the brink of the anxiety provoking.

Anxiety is an affect responding to loss. In particular, it's a response to the loss of loss itself. In other words, anxiety appears when where there should be a loss, there's instead something like an object. Loss applies to something that had been but is no longer. And from this we see that it can't be the lack of a loss. It isn't that loss was never there. Rather it's the case that there had been loss, but that loss is no longer. One is forced back into a wholeness that should not be. And here we get back to why the precious is on the brink of the anxiety provoking.

The precious, non-collectible collectible is an object that appears in the midst of loss. Something that is not isn't in the midst of anything. If something doesn't exist and never has, then we have no context for it. As in the quote from Benjamin above, there would be no "expressly devised historical system," for what is context but this historical system? The fact that the non-collectible collectible is one of a kind puts it in the context of a possible abundance. It creates this historical system, but can only fill it with one item. The non-collectible's collectibleness implies that there should be a number of these items to be gathered together, but the fact that there's only one imposes the constraint of *can* on that *should*. Thus, the one of a kind item is always in need of others. It becomes the most collectible for the very reason that the act of gathering together cannot be completed. The reasoning behind this being that if something is collectible and one has finished collecting, then upon finishing the act of collection, there's nothing "-ible" about it. The completed action of collecting results not in collectibility, but in the having of a collection. What we see, then, is that the most collectible item belongs to the incompleted, for there is no gathering together in the presence of only one. But what does it mean to be incomplete?

We can start to answer the question by referring to its opposite: completeness. That which is completed is that which, most generally, has come to an end. So completedness is always something that exists in a perfected state, whether past or present, or future perfect. It's something that's come to an end and remains having done so, or it's something that will have come to an end eventually. The incomplete, then, would *seem* to be that which could never exist in a perfected state. One might argue that there's no sense in which something can be "incompleted," for if incompletion were to become perfected, that would only mean that it would pass over into completion. But is this so? Is it possible that something can indeed be incompleted?

If we start something that has no possibility of finishing, then might we not say that its incompleteness exists in a perfected state? It is comparable to the reading of Benjamin's *Arcades Project*, for instance. We can never finish reading it because Benjamin never finished writing it. It is something that is perpetually incomplete, and it will remain so. After all, this is what it means to be perfected: for something to remain perpetually what it has become once it has passed into a certain state of existence. In this sense something can be incompleted.

Globalization is the same. We might speak of the post-global insofar as we have entered a stage in which the making of the world is frozen in this incompletion. Might we not say that to impose a structure upon something—a structure such as the one formed by the index of the network—is at the same time to impose upon that thing incompleteness? Structures are always formed around a negation, for things only come to be determinate through negation. But with the very structuring through negation, we are forced to deal with the negation of negation itself, and as such, this is that which cannot be absorbed by the structure. This is the very outside of structure that can never become determinate. And if all things that are determinate can come to be symbolized—if something is determinate, then we can have a name for it—then the negation of negation is that which resists being written, spoken, or gestured toward. The negation of negation would be everything as indeterminate. It would be everything as an indivisible something, for if negation is not, then nothing can be divided. Division only exists through negation. True, we can give a name to the state of affairs that exists because of the negation of negation. But this is only to name the state of affairs itself. What the negation of negation entails is that nothing *within* that state of affairs is namable. Whether that state of affairs is namable or not is inconsequential, for we are naming that state of affairs from the outside, not from within it. So, at bottom, what we are acknowledging is that when we impose a structure upon the world we introduce at the same time that which resists structure. This is the post-global. Put another way, when we make things determinate through the process of negation, we introduce the possibility of the negation of negation itself, something that leaves us with everything as indeterminate. Still, let's return to the library.

When we introduce an order to the library, we introduce along with it a boundary. I will show that this order isn't an ordering in the sense of arranging, an ordering in the sense of ordinality, in the sense of ordinal numbers, but in the sense of that which counts. This is an ordinality that implies that there is something to be counted. But this gets us into the spurious infinity. It's the infinite of the always one more. Rather, the negation of negation is the infinite, this thing that is incompleted. The idea of order makes it necessary for there to be a set that can be ordered and that that set contains more than one member. The members of this set are books. But the stipulation that there be more than one member to this set, a stipulation made necessary by the requirement of order, sets a limit. Though it doesn't impose a limit on the side of a progression—there can always be one more book—it does impose a limit on the side of inception. There's a limit on the side that one must begin with at least two of something. Thus, the first of the set is equal to two. Order cannot be made within a set wherein division is not possible, in a set that consists of one and only one. Thus, any ordered library is finite. And, again, I do not mean by *order* a system of organization. It isn't as though alphabetizing or imposing a system of categorization by subject matter makes a library finite. Rather, what I'm suggesting is that the requirement that order imposes of there being at least two is what makes a library finite. Thus, an infinite library that's always infinite in and of itself—infinite without partaking in the spurious infinity of always one more—is one that isn't bound to the rule of divisibility. An infinite library is one wherein division does not exist. An infinite library is a set that consists always only of one. But what would this mean? It would mean that we don't think about the library as a set of books, but as a set that contains one member. It is at this point that we can return to our question about the extinction of the book.

In "School Library," Benjamin describes his experience of the assigned readers as this: "I had to remain confined within particular stories, as if within barracks that—even before the title page—bore a number over the doorway." (Benjamin, 2006, p. 143) There's something binding about this description of the book, and we might supplement this idea with what Jacques Derrida puts forth in *Paper Machine*. It's the idea of a gathering together that's the book's most salient quality. A book is a gathering together under a title, a title bearing the book's name, its identity, its legitimacy, its copyright. This gathering together extends to the library, the gathering together of a store of books (Derrida, 2005, pp. 6–7). But, Derrida asks, might we still call a library that which no longer gathers together this store of books? For what would be a gathering together of texts with no paper support, electronic texts that are not finite, that may be open textual processes over global networks, texts that may be actively or interactively co-authored by readers? Could it be that the book as we know it has no future, or that if it has a future, it will no longer be what it was? Could it be that we're awaiting another book that will rescue the book from what is now happening? And what of this book to come, a book whose past has not yet reached us, but whose past we nonetheless must think about? Might we not say that the library of the network is exactly what is this book to come? Ideally, it's a gathering together of the totality of all books that at the same time creates a dispersion, a dispersion because its searchability has no respect for titles, or that which is the delimiter of the book *qua* book.

So here we have thus far had two descriptions of books, one from Benjamin and the other from Benjamin as supplemented by Derrida. The book is that tactile object that one can possess and at the same time an abstract something gathered together under a title. With both descriptions, the book can be collected in a library, but one is the private library of a collector, a library that can be unpacked. In the other description, the book belongs to a public library that has the potential to contain everything in a way that can't be unpacked. In the literal sense, the library of network can't be unpacked. It's a unified whole; other than orthographical symbols, there are no discrete units as far as the searching is concerned. Figuratively, this library belongs to no one in particular, so there's no one to do the unpacking. The book collector ceases to exist, becomes extinct as Benjamin might say, and this is so not only because there is no need to gather together books, but because the library has moved away from the model in which books are members of a set that constitutes the library. Rather, through the coming of the library of the network, the conception of the library has shifted to the *totality of the published*, to the totality of publicly readable text. It is precisely in this way that the library becomes infinite. It becomes a set that can only have one member. It is infinite insofar as that one member necessarily no longer retains divisions. True, the member of that set is in perpetual flux. It does indeed change every time more text is added to it. Still, it remains the totality of all that has been published, only it gets bigger with each indexed addition. But we again return to the book collector. I have explained collecting in general, but who is this *threatened* collector?

V

Though the practice of intellectual inquiry is not limited to the scholar, it's the labor most appropriate for the scholar. It stands to reason, then, that it's the scholar who is most affected by a change in the means of intellectual inquiry, a change that will come from the existence of a totalizing and searchable library. For the scholar, this would mean that a large part of scholarly labor will become migrated to the machine. Interfacing with a digital library through a search engine would allow us to access knowledge in a different way. It isn't merely that it allows space to be traversed with greater speed, but memory becomes a prosthetic memory. If anyone can be as smart as they are able to use a search engine, might this be the extinction of the scholar? Will the intellectual be anyone with access to the network? Is this how we will comprehend the scholar, only after the scholar becomes outmoded? But perhaps the scholar is not the same as this collector. Consider what Benjamin says of the historian Eduard Fuchs:

> From the outset, he was not meant to be a scholar. Nor did he ever become a scholarly 'type,' despite the great learning that informs his later work. His efforts constantly extended beyond the horizon of the researcher (Benjamin, 2002b, p. 263).

Perhaps the scholar can evade extinction by *becoming* a collector. Perhaps it's the case that the scholar who is threatened with extinction wasn't a collector to begin with. Let's examine the ways scholars interface with both the non-searchable library and the infinite library of the network. This will give us some clarity.

The way scholars interface with the non-searchable library involves a certain amount of necessary ignorance. Many of the books we've read we've read without having known whether or not they'll prove to be useful. We finish books with the hope that they'll contain something of use, but we can never tell until we finish them. Consequently, we end up finishing many books that we don't find immediately useful, but still, we consider ourselves better off for having read them—unless of course the book was bad, but in that case it's likely that we wouldn't have finished it. Further, we happen upon many of these books by accident. They have interesting titles, or the dust jackets make the book seem promising, or the book is from a press we like. Or perhaps we've read a book review, or the book was cited in another book, or we've been recommended that book by an acquaintance. In all cases such as these, we read the book with only a vague promise of usefulness. We must interface with books in a way that doesn't allow us to know what we're missing—or not missing—until we've read them. Our assessment of the book's utility only comes after having finished it. If we start interfacing with the library via the efficient network, then things become much different.

With a networked library, we start with an interest, search that interest, and all the relevant networked objects come to be at our disposal inasmuch as we're able to access their locations. This is so, for we're searching more than just subject headings and titles. We're searching within the text itself. Instead of having a vague promise of usefulness, we start with a computer-mediated assessment of a text's utility, and once we've decided upon its utility for ourselves, we proceed to read that text. The technology of the search engine, an incarnation of the concordance of the efficient network, seems to change our reading practices. Now, as scholars, we generally try to read everything that's out there so we can have found everything that's suited to our interest, and our interests often become shaped by our attempts to read everything we're able.

Still, the goal should be to prune away the inessential, to establish a collection of the useful, to make our personal canon. Consider what Benjamin writes about the book in *One Way Street*.

> And today the book is already, as the present mode of scholarly production demonstrates, an outdated mediation between two different filing systems. For everything that matters is to be found in the card box of the researcher who wrote it, and the scholar studying it assimilates it into his own card index (Benjamin, 2004, p. 456).

This is what the scholar as collector accomplishes. The scholar who is also a collector makes such a "card box" and transcends the book. If the book is but an outdated mediation, perhaps the mediation of the infinite library of the network should remain precisely that: a mediation. The scholar needs not face extinction because the network is but an in-between. The network doesn't threaten to wipe out the scholar. Rather, it facilitates the creation of scholars.

But this isn't to say that the network is a genie granting us genius. Genius must be protected. Certain uses of the network can threaten genius, in fact.

Less is left to accident with the use of the library of the network, and the glut of what our searches return will leave little time to read anything else, for if we're still operating with the imperative of scholarly rigor—with the imperative of knowing everything that's out there in our area—we become answerable to the question, Why haven't you read this? We have no excuse not to have known about something if knowing about it only involves a search engine search. We'll end up having less tolerance for ignorance—in both ourselves and others—but this intolerance will only enforce a self-perpetuating focus upon the interest with which we've begun.

With the imperative of rigor, to know all that which has already been known before proceeding in my goal to create something else to be known, then I am prohibited access to the unknown by the searchable library of all texts. In this case, I'm confronted by the weight of a historical malady. In the case of the non-searchable library, if there's something I ought to do, I must first be able to do it. Thus, the imperative of rigor implies that I must know all there is to know about my object of inquiry before I proceed. However, I can't know that I know all there is to know without having read all texts. Thus, I can only approach the ideal of rigor because I have no means by which to realize rigor in any practical way. Whereas I would've had to have read all texts in order to know what I had not known, the searchable networked library allows me to proceed efficiently. It makes knowing all there is to know about my object of inquiry within the realm of the possible. In other words, I don't have to read the totality of all texts, only texts that are relevant to my object of inquiry. This will often prove to be a nearly impossible task, but it isn't the absolutely impossible task of reading everything.

Mastery becomes impossible because the search technology makes it theoretically possible. Mostly, there will still be too much to read, only we'll always know what we're missing because all of the citations will have come up in the results of a search engine query. Without search engine technology, however, mastery is possible because we go into the activity of attempting mastery with the foregone conclusion that mastery is always only possible. Here, though we never end up mastering that which was always only possible anyway, we still accomplish a kind of mastery, albeit a different kind. We become well read.

With the networked library, all knowledge exists as an accessible totality at once, and although this has always been the case, it hasn't been practicably accessible because of ignorance. Ignorance as a mediator doesn't usually vanish. Not knowing that a text exists doesn't allow for a true synchrony. However, with the networked library, ignorance does tend to vanish. Knowledge becomes immediate insofar as it's no longer mediated by ignorance.

So what we might see, then, is that genius has thus far proceeded from ignorance and aimlessness. Again, we might find resonances with this in a quotation from *The Arcades Project*:

> Basic to flanerie, among other things is the idea that the fruits of idleness are more precious than the fruits of labor...Most men of genius were great flaneurs—but industrious, productive flaneurs (Benjamin, 2002a, p. 453).

Like the strolling flaneur, we must read industriously, but resist the mechanistic systemization that can come with certain usages of the infinite, networked library. We must continue to be scholars, but resist that position as a position resulting from the division of labor. True, we are subjected to this labor and may find ourselves alienated in it by the requirements of rigor, but we must approach intellectual inquiry as a labor of love. We may be able to enjoy a certain amount of ease with the infinite library of the network, but this is an enjoyment we must refuse. Let's return to that vignette I've been reading over and over, the one titled "Boys' Books." Here's the beginning of the ending paragraph:

> Or is it with older, irrecoverable volumes that my heart has kept faith? With those marvelous ones, that is, which were given me to revisit only once, in a dream? What were they called? I knew only that it was those long-vanished volumes that I had never been able to find again (Benjamin, 2006, p. 60).

There's something about keeping faith with that to which we must have a relation of not knowing. Certain usages of the infinite networked library threaten to rob us of this not knowing. It is not so that because we can enjoy something, we must. For it's an enjoyment of ease afforded to us by the infinite library of the network that steals the enjoyment of discovery, a discovering enjoyment that reproduces itself, an enjoyment like that of when we were children reading at a time when we "still made up stories in bed." (Benjamin, 2004, p. 464) What is lost is a loss similar to the "I don't know" of childhood. We might try to retain an enjoyment of the library similar to that of what Benjamin describes of Karl Hobrecker, a collector of children's books: "That childlike pleasure is the origin of his library, and every such collection must have something of the same spirit if it is to thrive." (Benjamin, 2004, p. 406)

As we've seen, the most collectible items aren't collectible in the sense that we can gather them together. The most collectible are one of a kind. With the library of the network, we already know beforehand that an abundance is lost. When everything is unveiled with the searchability of the infinite library of the network, what's lost in abundance is the ignorance of not knowing what's out there. What's lost is the abundance of potential discovery. What becomes most collectible, what becomes one of a kind, then, are our strolls as scholars, as intellectuals. For there's still something that can't be gathered together by the index of the network, and that's the particular path we stroll. A stroll is movement, and because movement isn't divisible, it can't be gathered. It can't, in other words, be indexed. Strolling resists the network. We must now learn how to move boundlessly in this way through the boundless networked library. It's only by means of such singular movement through knowledge that we can have a relationship to the oppressive wholeness of totalized knowledge and remain scholars.

To have mapped the world is not to move through it. To apprehend the globe with the view of the gods is not to see it as the human traveling on foot. This is what must remain. We cannot allow the network to steal our intellectual wanderings. The infinite library of the network should merely give us a bigger city, not an itinerary.

REFERENCES

Benjamin, W. (2006). *Berlin Childhood around 1900*. (H. Eiland, Trans.) Cambridge, MA: Harvard UP.

Benjamin, W. (2004). *Selected Writings, Volume 1, 1913–1926*. (M. W. Jennings, Ed.) Cambridge, MA: Harvard UP.

Benjamin, W. (2005). *Selected Writings, Volume 2, part 2, 1931–1934*. (M. W. Jennings, Ed.) Cambridge, MA: Harvard UP.

Benjamin, W. (2002b). *Selected Writings, Volume 3, 1935–1938*. (M. W. Jennings, Ed.) Cambridge, MA: Harvard UP.

Benjamin, W. (2002a). *The Arcades Project*. (H. Eiland, & K. McLaughlin, Trans.) Cambridge, MA: Harvard UP.

Derrida, J. (2005). *Paper Machine*. (R. Bowlby, Trans.) Stanford: Stanford UP.

Hegel, G. W. (1969). *Science of Logic*. (A. V. Miller, Trans.) London: Allen & Unwin.

Nancy, J. (2007). *The Creation of the World or Globalization*. (F. Raffoul, & D. Pettigrew, Trans.) Albany: SUNY Press.

CHAPTER THREE

Travel in the Network

Redirected Gazes, Ubiquitous Connections and New Frontiers

ULRIKE GRETZEL

INTRODUCTION

Technologies not only enable travel but also fundamentally structure the experiences travel brings about. Travel involves movement through geographic space and time, and technology transforms the relationship between the traveler and the traveled space as well as the experience of time. For instance, Schivelbusch (1986) describes how the railroad turned travelers into projectiles being shot through the landscape, leading to a loss of sensory immersion and a particular perception of landscapes as distant, evanescent panoramic views. Nowadays, emerging information and communication technologies add to the mediation and mediatization of tourism experiences. Jansson (2006, p.1) stresses that "mediatization alters perceptions, of place, distance, sociality, authenticity, and other pre-understandings that frame tourism." Much has been written about time-space compression in the post-modern society (Harvey, 1989). In the network society, these effects are amplified by network technologies to the extent that individuals experience time as timeless and space as placeless (Castells, 1996). This of course has fundamental implications for an activity like travel that is very much determined by these two dimensions.

Access to network technologies changes the travel experience itself in that notions of home and away become blurred (Rosh-White, & White, 2007). Memories of a trip that is not over yet can be instantly shared through the technology and messages from home or work can easily interrupt experiences at the destination (Jansson, 2007). Friends and family at home can follow every move of the traveler if updates are posted online. Internet cafes have become part of the touristic infrastructure and Internet connectivity is an important amenity offered by hotels and sought after by travelers to be able to stay connected while being away. Digital cameras, now very often integrated in mobile technologies, allow for almost unlimited picture-taking, instantaneous review of what has been experienced, and sharing of experiences while a trip is still ongoing. New media, especially smart phones which include tools such as navigation systems, can "eliminate some of the socio-cultural friction of touristic mobility" and affect travelers' appropriation of foreign terrains (Jansson, 2007, p. 13) in that they guide tourists to their locations and render the asking of questions regarding directions unnecessary, provide suggestions for restaurants or deliver interpretive contents through, e.g., podcasts. However, technologies also create new opportunities for social interactions. Travelers can connect with other travelers and with locals through social networking tools such as WAYN.com and Couchsurfing.com. Technology also changes the experience at historic sites, attractions and museums in that it is increasingly used for the representation of objects and sites, and allows travelers to be more actively engaged with contents that could even be personalized to fit their needs. Overall, new media technologies can provide "more detailed scripts of potential journeys; aiding tourists to coordinate their touristic activities more efficiently; making touristic representations more negotiable and ready for immediate transmission/sharing" (Jansson, 2006, p. 29).

While tourism is facilitated by many types of technologies (Werthner & Klein, 1999), network technologies play a particularly significant role. Both the Internet and mobile technologies have been identified as especially important agents of change in the tourism industry (Buhalis & Law, 2008). Mobile technologies play an increasingly important role in tourism due to their ability to provide travelers with wireless and, thus, instantaneous and pervasive Internet access. Handheld devices such as personal digital assistants (PDAs) and cellular phones supported through a wireless application protocol (WAP), a global system for mobile communication (GSM), and short message service (SMS) allow travelers to take full advantage of the Internet while on the road. More ambitious developments of mobile technology go beyond simple access by providing real-time location-based services (Eriksson, 2002; Oertel, Steinmüller, & Kuom, 2002). Empowered by geographical information systems (GIS) and global positioning system technology (GPS) in combination with information available on the Internet, these advanced mobile applications identify the traveler's location in space and the spatial context of this position. This information is then used to generate personalized assistance in the form of location-specific and time-sensitive information. Many advancements related to mobile technologies are spurred by needs that directly arise from information and communication problems encountered during travel. CRUMPET—creation of user-friendly mobile services personalized for tourism (Poslad, Laamanen, et al., 2001; Schmidt-Belz, Makelainen, et al., 2002)—and the development of wireless-based tourism infrastructures, for instance the ambient intelligence landscape described by the Information Society Technologies Advisory Group [ISTAG] (2002), are only two examples of the many efforts undertaken at the juncture of mobile technology and tourism; yet they represent developments with important implications for the future of tourism and the nature of what constitutes tourism experiences.

This chapter examines changes in the very notion of travel brought about by Web-enabled mobile phones such as the iPhone. A recent study shows that 40 percent of worldwide Internet users used a Web-enabled phone, with iPhones clearly dominating the market followed by BlackBerries (eMarketer, 2009a). These devices allow travelers to access and distribute information through both the mobile communication network and the Internet. They constitute an important link between the Internet and mobile communication networks. Consequently, they exaggerate many of the aspects that can be found in the two network types and warrant careful examination in the context of travel. Specifically, the chapter seeks to establish a connection between the technological components of the network and the "network culture" (Terranova, 2004) as it pertains to travel.

CONCEPTUALIZING THE NETWORK

Networks are simultaneously material and immaterial (Galloway & Thacker, 2007). Underlying the assumptions in this chapter is a conceptualization of the network as having various layers beyond the technological infrastructure, including social, political, cultural, psychological and emotional dimensions. In the context of Web-enabled phones, the material aspects of the network include the cell sites (often referred to as cell towers), the mobile phone devices, SIM cards, cellular telephone exchanges (switches), satellites to enable GPS functions, as well as the hardware and software components that make up the Internet. While mobile communication networks are aggressively expanding their coverage and the Internet continuously grows in terms of server nodes connected to it and content accessible through it, there are clear limitations in terms of where services are accessible to Web-enabled mobile phone users. Such a physical network requires, of course, technical coordination at a global level. Both the Internet and mobile communication networks would not be conceivable without the implementation of technical standards. However, globalization in the network is not limited to the technological infrastructure but is also driven by the speed of cultural and informational flows (Terranova, 2004). The widespread technical coordination reflects new levels of social coordination, thus hinting at the social and political dimensions of the network and providing what Grewal calls a striking example of "network power in action" (2008, p. 14). Coordination is also required at the organizational level, best exemplified by so-called roaming agreements among mobile phone providers. Organizational coordination, or rather the lack thereof, is especially visible to the traveler who might leave her "home network" and need service in a "visited network."[1] Recognizing mobility within the network and the increasing reliance of travelers on mobile technologies, the European Union has regulated roaming rates, including a recent introduction of caps on text message roaming fees. Coordination is both liberating and entrapping in that it leads to greater access but also privileges certain modes of access over others (Grewal, 2008). Thus, coordination in the network is inherently linked to issues of power and control. In the end, power and powerlessness in a network context are a function of access to the network and the control of flows (Barney, 2004). Galloway and Thacker (2007) stress that networks exercise novel forms of control at a level that is anonymous and non-human; a form of control that is embedded in the technological infrastructure similar to that of the panopticon (Foucault, 1995). However, control is not exclusively embedded in the technological dimension of the network. Control in the context of Web-enabled phones, for instance, is indeed an issue of access and technological affordances, but also, among others, an issue of contractual ties between device manufacturers and mobile service providers, authority over contents and applications, active rather than just passive tracking of users, the creation of dependencies through multi-year contractual usage agreements, and marketing that makes the technology accessible to some but not others and promotes certain uses. Control is further visible in the social construction of these technologies as indispensable and life outside mobile Internet access as scary and utterly undesirable.

Importantly, networks are not just tools but rather form an important part of our collective consciousness (Galloway & Thacker, 2007). They infiltrate our values, our language and our social interactions. Being a network technology user can become an essential signifier of identity (Turkle, 1995). Not having a Web-enabled mobile phone can mean social exclusion. A recent Pew Research Center report argues that mobile technologies deepen the digital divide, increasing the cost of being excluded from or only engaging at low levels with the network of people and information found online (Pew Research, 2009). In addition, like other media, network technologies can change our perception, cognition, and our memory (McLuhan, 1964; Grossberg, Wartella & Whitney, 1998). Clough (2008) describes new media as providing the possibility of a profound extension of our senses. Network technologies can even become physical extensions of our bodies as in the case of Bluetooth headsets attached to the ears of mobile phone users. As such, digital technologies "attach to and expand the informational substrate of bodily matter and matter generally" (Clough, 2008, p. 2) and influence the body's capacity to experience the passing of time (Hansen, 2006). The network also has an affective dimension that is often overlooked (Clough, 2008). Using the network or being excluded from access to it can invoke emotions across a wide spectrum, ranging from content and joy to the surprise of being disconnected, frustration and fear. The following provides a discussion of Web-enabled mobile phones in the context of travel, taking into account the technological as well as the political, social, cultural, psychological and affective layers of the network.

GAZING AT AND THROUGH THE WEB-ENABLED MOBILE PHONE

Much has been written about the tourist gaze (Urry, 2002a). Based on Foucault's notion of an institutionalized way of looking, Urry defines the tourist gaze as the organized and systematized way in which tourists look upon landscapes, natives, historical sites and other objects of interest. The tourist gaze is often directed by tourist guides in the case of organized group travel or by guidebooks such as Lonely Planet in the case of individual travel. Tourism marketers have a stake in directing it successfully to their own offerings. Interpretive panels, pamphlets or audio tours provide tourists with further interpretive information to help them make sense of what they are gazing at. These sources reinforce certain images and promote dominant views, thus influencing tourists' perceptions of what they look at and what they should be experiencing while at a destination (McGregor, 2000; Bhattacharyya, 1997; Koshar, 1998). The gaze can also be mediated by technologies such as binoculars, coin-operated telescopes, video-cameras and photo cameras. Rather than taking in the sensory information through one's eyes, tourists use technologies to extend their sensory reach. Thus, the technologies used by tourists to gaze at objects frame the view, change patterns of perceptions (McLuhan, 1964), and also create a barrier between the tourists and the objects of their gaze. The result is a very particular mode of experiencing. The tourist gaze has been criticized as a particular way of seeing that is very passive, superficial, condescending, and often leading to commodified experiences (McGregor, 2000). While distant and superficial gazing is not necessarily a dominant activity for all tourists (Perkins & Thorns, 2001), it is a very common way for tourists to approach and experience "the exotic."

It is argued here that the tourist gaze is increasingly controlled by the network. The authority of directing the gaze is taken away from the traditional authorities such as tour guides and visitor information centers and now rests with the mobile device and its underlying communication network. Web-enabled mobile phones redirect the gaze in that they help tourists rely less on touristic canons published in guidebooks or preached by tour guides. Interpretive facts can be checked, alternative sources can be consulted, consumer-generated podcasts can be downloaded, and additional, personally-relevant information can be accessed easily while gazing at a site. At the same time, it is questionable how reflective tourists will be when using these resources. Also, many of the sources still rely on the very same editorial contents promoted through traditional guidebooks. Rather than using the print versions of the tourist guidebooks, the beaten paths are now laid out by the information that is accessed online. The web-enabled mobile phone becomes the new "marker" (MacCannell, 1999), pointing out what should be considered as an attraction, powered by the applications and contents it features. Instead of diffusing authority, it actually creates new levels of authority and control while giving the impression of freedom and choice. Theoretically, there is indeed a greater opportunity for individualization and gazing beyond the beaten path given the number of sources and applications that are available. Practically, there will be preferred applications, featured contents, and new advertising models, providing new opportunities for authority over the tourist gaze to be exercised.

The tourist gaze is typically communal in that people in the same travel party or travel group experience the same things and form strong social bonds based on such shared travel experiences. Staying together was an imperative before mobile phones became widely available as it was often complicated to meet up again in an unfamiliar environment once the group split. Traveling together meant that time, space, and objects of the gaze were shared. Travel within the network allows for more individualization in the gaze as one is able to explore destinations and attractions on one's own. It is easy to part and then convene again as a group when messages can be exchanged. Thus, travel experiences in the network might not be shared in the sense of being in the exact same place at the same time but rather through information exchange during and after the individual experiences occur. It also shows how the network penetrates social facets, potentially changing the social structure of travel groups.

A lot of the directing of the gaze is done through means of specific applications, which are small programs that can be downloaded to the iPhone. Apple estimated in January 2009 that 15,000 such applications were available and that it had served 500 million application downloads (eMarketer, 2009b). The iPhone currently features 242 Web applications specifically related to travel. These applications range from more generic searches for "Things to Do" in different cities to very eclectic applications, for instance travel maps for Chinese Kung Fu fans and information about liquor stores in Madison, WI provided by I'mSoBombed.com. Personalization of the travel experience is emphasized in many of these applications. Referring to the ability to become independent of other tourism mediators, the "Pocket Norwich" application stresses:

> The guide is unique in that it includes full audio commentary, streamed directly to your device—let your mobile become your own personal tour guide!

Similarly, the "EyeTour Puerto Rico" application promises:

> You are now ready to Be Your Own Guide!

Being able to draw on the knowledge of a community of travelers and of locals instead of having to rely on marketer and guidebook information or having to actually ask locals at the destination, is also a rationale used for several of the applications. For example, the "hiogi.answers.everywhere." application, which allows travelers to type in questions and receive immediate feedback, promotes itself using the following argument:

> The most convenient source for local insider knowledge and travel guide information. The hiogi-community knows everything and so should you.

Not only the gaze but also communication can be mediated by these applications. Language translators are becoming ever more popular and are also available for the iPhone. For example, the "Lastminute.com Talking Translator" application provides audio translations of common phrases in several languages. More sophisticated translation applications for mobile phones are currently being developed and promise instantaneous speech recognition and translation in the form of voice outputs (Watanabe et al., 2000; Paul et al., 2008). Other applications are focused on more functional aspects of the tourism experience, such as hotel deals and bookings, car rentals, restaurant reservations, currency conversion, train schedules, airline arrivals and departures, and even private jet charters. In general, there is a clear emphasis on supporting the planning aspects of tourism, leaving very little room for exploratory gazes. Stimulation in the sense of exploring the unknown, having unpredictable experiences, and experiencing risks is an important motivation for travel identified in the traditional tourism literature (Pearce, 2005). However, the great number of iPhone applications that stress efficiency and advance planning appear to suggest that exploration is a wasteful way of experiencing a destination. Indeed, the "iconFind" application condemns the purposeless wandering around and taking in sights as one stumbles across them. It firmly states:

> Don't wander around: use iconFind today!

Similarly, the Vegas.com Mobile Concierge application demands focus and immobility:

> Find shows and tours, read reviews, book tickets, and more—without ever having to leave the casino or the pool. It's the way Vegas was meant to be done!

The applications further provide new opportunities for gazing. The "Motion Life Travel" application uses "being able to see your destination on Travel Television" as one of its main selling points. Consequently, the tourist gaze can already start long before one reaches the travel destination. There are also several applications for Webcams, e.g., for viewing what happens on Princess Cruiseline ships or in New York City. In this sense, the web-enabled phone becomes a new, mobile "panoptic device" (Foucault, 1995), as one can now investigate what other tourists do from anywhere, without having to be at the destination; and if one is at the destination, one has to assume that others will be watching. These applications also support new forms of "armchair travel," allowing for touristic gaze to happen wherever the mobile phone can be used. "WorldPics," for example, praises the opportunity to explore the world without leaving home:

> *WorldPics.fr is a free community website made to discover the world through geolocated photos uploaded by the members. Take part to the adventure and visit the world in your couch. Enjoy!*

However, the Web-enabled mobile phone also redirects in the sense of focusing the gaze away from one's environment and onto the device. The screen becomes the new object of the tourist gaze. It is increasingly common to see people on trains, buses or walking down a street looking at their device rather than their surroundings. One can now be engaged with contents and activities afforded by the phone. No need to look around and savor new environments when the object of one's fascination lies in one's palm. For instance, one does not have to engage with the real San Juan as "EyeTour Puerto Rico" lets one watch videos of the city instead. Also no need to interact with locals or fellow travelers if the phone provides the necessary diversion and entertainment. Thus, web-enabled mobile phones can promote superficial tourism experiences in that they make it very easy for travelers to disengage with the actual surroundings. Experiences of time and space as essentially limiting elicit a "sense of place," but such localized experiences become nearly obliterated in the network society (Barney, 2004). Travelers no longer obtain a deep sense of place when experiencing the destination through the network. As a result, localities become disembodied from their cultural, historic and geographic meaning, creating a culture of real virtuality (Castells, 1996) that infiltrates travel experiences and at the same time alters the meaning and dynamic of places.

These specific examples clearly illustrate that the layers are tightly interwoven. Redirecting the gaze in a technological sense changes perception and results in a redirection in a cultural sense as well. They change the social fabric of travel and either heighten or discourage emotional engagement with what is actually experienced. The examples also show that issues of control and authority are pertinent, even if some aspects of the network are open and collaborative in nature.

TRAVEL WITHOUT EVER LEAVING HOME BEHIND

Travel is marked by the distinction of "being at home" and "going away" (Jansson, 2002). However, the network "forms and deforms the fabric of time and space" (Terranova, 2004, p. 40), making it increasingly difficult to draw clear boundaries between home and away. Travel is synonymous to leaving the home behind. Thus, the concept of "home" is an important element of people's construction of what it means to be away (McCabe, 2002). Home is not only a physical or geographical concept but also comprises emotional and social relationships (Rosh-White & White, 2007). Jansson defines tourism as "a kind of organized retreat from the temporal and spatial features of labour practices and everydayness" (2002, p. 431). In fact, escaping the everyday-life is a motivating factor for many travelers (Ryan, 2002; Ateljevic & Doorne, 2000), as are the avoidance of interpersonal stress and getting away from one's family (Pearce, 1995). It is not clear how these needs can be satisfied when one remains in constant contact with those at home. Instead of being rooted in the time and space of the destination, one is actively keeping track of what time it is at home to know when to contact family, friends or the office and remains connected to various localities at the same time through news updates, photos, etc. Yet, on the other hand, such communication can provide much needed emotional support, possibly decreased feelings of guilt for leaving others behind, and an opportunity to build social status by being able to instantaneously share experiences with others. Maintaining relationships with family at-a-distance is common. Traditionally, the separation from family due to travel was managed with telegrams, letters and postcards while today, mobile phones and the Internet have significantly increased the ease of keeping in touch (Rosh-White & White, 2007). Beyond the basic call and texting capabilities of mobile phones, web-based applications make it increasingly easy to stay in touch with those at home. For instance, one can access Facebook through one's iPhone and broadcast status updates and newly uploaded photos and comments to others in one's social network. "TripBounce" is an application that allows for instantaneous updates of one's blog. If one would like to stick to traditional modes of communication while taking advantage of the convenience of the mobile phone, "Wish You Were Here (SM) Cameraphone Postcards" is an application that allows one to send pictures taken on the mobile phone as postcards to any address in the US. However, the communication is two-way in that family and friends at home can also communicate with the traveler, sending them updates about the current affairs at home. Thus, the connections with home remain largely intact while traveling. Rosh-White and White (2007) found in their interviews of international campers in New Zealand that the establishment of co-presence with family and friends at home was integral to the travel experience and that only a small minority of travelers truly tried to break away from contact with those at home. The sense of connection with "home" seems to be important and the plethora of iPhone applications available make it easy to establish at least symbolic proximity with others (Wurtzel & Turner, 1977). In this sense, travel in the network is very much characterized by virtuality and involves experiencing timeless time and placeless space (Castells, 1996) as one is simultaneously active in different time zones and places.

Phones historically played an important role in the promotion of tourism as they allowed factory owners and other important businessmen to leave work without being out of reach in case of emergencies that required their immediate attention. Indeed, some of the earliest phone lines were installed in hotels to serve this clientele (Aronson, 1977). Business travelers without mobile phones is an almost unimaginable concept nowadays. They reach out and can be reached wherever they go and frequently conduct business using web-enabled mobile phones while being away from their offices. From the perspective of tourism, this ability to be reached whenever and wherever one travels constitutes an interesting development in that it blurs the boundaries between work and leisure or vacation and makes escaping the everyday-life an ever greater challenge. Traveling within the network leads to expectations that one checks messages and responds to requests on a continuous basis. Airplanes are currently a mostly "network-free" zone when in the air; however, several airlines are experimenting with inflight Internet access and it is only a matter of time until mobile phone use during the flight will be a common activity. The psychological stress this need to disconnect for the flight creates is very visible in the fact that flight attendants have to enforce the obligatory switch-off of phones and other wireless devices and many passengers turn their devices on again the very second the plane touches the ground.

Travel cannot only be motivated by escapism but also by social needs. Ryan (2002) stresses that social motivations include the need for friendship and interpersonal relationships as well as the need for the esteem of others. Social interactions in the context of travel have been conceptualized as occurring between tourists and other tourists (including their immediate travel party), tourists and service personnel, and tourists and locals (Pearce, 1995). Clearly, communication with those who are at home or in other parts of the world have not been considered in these theoretical models of the social fabric of travel. These previous conceptualizations of social interactions in the context of travel also typically assume that social interactions with the above-mentioned groups happen face-to-face rather than through communication media. It seems that traveling with a web-enabled mobile phone, having many applications at hand which allow for communication and social networking, tremendously impact opportunities for social interactions.

On one hand, the web-enabled mobile phone opens up new avenues for communication and for having social experiences while traveling. Several new applications build on the desire to meet others who happen to be in the same place at the same time. Urry (2003) explains that such technologies are necessary in a network society where connections are spatially dispersed and membership of one network does not necessarily overlap with that of any other, as this lack of overlap reduces the likelihood for quick, casual meetings and chance encounters. For instance, the iPhone application "Travel Buddies On the Go" allows travelers to share travel plans and trip thoughts while being on the road. If travel plans overlap with those of others, this is easily visible through the application, and meetings or joint travel can be more easily arranged. "Mobile Map Me" is another application which supports social interaction through the mapping of location-based data. It promises travelers the ability to create their own maps, connect with friends by seeing where they are, share location and status updates for others to see, and explore places recommended by friends. "Loopt" is an application available on the iPhone that enables friends to share location-based information. Specifically, "Loopt" allows users to make their whereabouts known to a select group of friends so that impromptu gatherings are possible if two friends are in the same vicinity (Sharma & Vascellaro, 2008). "Google Latitude," available through "Google Maps for mobile," is another mobile phone-based application that allows users to update their location on the go and see where friends are if they are willing to share their location details. It integrates "Google Talk," allowing users to update status messages and profile photos. It is not yet available on the iPhone. In general, these applications allow travelers to "leave traces of their selves in informational space" (Urry, 2002b, p. 266), increasing possibilities for social connections and for feelings of co-presence (Rosh-White & White, 2007).

On the other hand, the use of mobile phones while traveling and the need and/or desire to maintain communication with those at home can interfere with communication with others while traveling. Social interactions are oftentimes an integral part of travel experiences. Direct communication is an important aspect of such social interactions. For Heidegger, talking is the fundamental activity by which people express their experience of being with each other (Myerson, 2001). While using the mobile phone to communicate with family and friends can allow for the sharing of experiences, it reduces the opportunities for shared experiences with those at the destination. If time is spent communicating with those at home, there might not be any time or desire left to interact with fellow travelers or locals. Thus, the social aspect of tourism experiences might get lost. Experiencing the "Other," the "Exotic" is an important driver for tourists (Ryan, 2002), yet being immersed in maintaining contact with home and nurturing existing relationships might make it impossible to fulfill these needs to the extent that they are fulfilled when communication is focused on those at the destination. Communication sometimes does not even happen between those in the same travel party, as exemplified by family members texting or talking on the phone while sitting at the same table in a restaurant at the destination.

REAL ADVENTURE IS WHERE THE MOBILE PHONE WON'T WORK

While some travelers seek out comfort, others are highly motivated by the prospect of adventure (Pearce, 2005). The adventure traveler seeks new experiences in new environments primarily for the sake of personal challenge and arousal (Jansson, 2006). Adventure travel involves risks and pushing limits. The very notion of adventure implies that help cannot easily be summoned, that one has to rely on one's own skills or on the support of the group one travels with. Adventure also implies that information is scarce or unreliable. Adventure mostly happens in areas that are defined as wilderness, areas which are distinct from developed, "civilized" places. These regions are also often called "remote" or "peripheral." Their remoteness is usually based on their inaccessibility by common means of transportation, their lack of connections to the highway or railroad networks. These physical networks are severely hindered in their expansion by obstacles such as mountain ranges, rivers, etc. and involve considerable investments. Wireless networks which transmit digital signals do not face the same barriers. Antennas are less costly and can be strategically placed to provide seamless coverage in areas that cannot easily be accessed. Still, some areas are not covered, as low phone usage in these areas makes the installation of antennas cost-inefficient. Areas that are connected become distinctly different from those that are not. The network transforms what it connects and leads to new geopolitical formations (Terranova, 2004). The territory of the wireless network is marked by cell phone towers and the new frontiers lay beyond the signal reach. While "inside the networks, new possibilities are relentlessly created—outside the networks, survival is increasingly difficult" (Castells, 1996, p. 171).

The mobile phone provides a ubiquitous safety net, as long as one travels within areas where network coverage exists. Calls for help can be placed, information can be accessed, one's location can be known and communicated to others. Traveling to an area where one's mobile phone will not work creates anxiety. This anxiety is not only caused by the realization that help cannot be accessed but often also by a fear that one misses out on important news, that one cannot connect with loved ones, and that one actually has to engage with the unfamiliar. Thus, the new frontiers of adventure are now defined by network availability. Adventure is no longer marked by treasure maps but by coverage maps. Danger lies in not being able to use one's web-enabled mobile phone. A recent snippet in Outside Magazine (2009, p. 82) illustrates the point, describing adventure as only being possible outside the technical reach of the network:

> Wilderness Wisdom—The Voice of Dissent. Doug Peacock—the model for Ed Abbey's Monkey Wrench Gang antihero, George Hayduke—tells you how to get lost, go deep, and have a real adventure…."Forget all the risk-free literature you read from the BLM or the Park Service or anybody else that's covering their ass legally. And no GPS devices; the opportunity to get lost on today's planet is a privilege."

Places without network coverage are called "dead zones," implying that something very negative could happen if one enters these areas. The "scariness" of travel beyond the network is well-illustrated in the Verizon Wireless "Dead Zone" commercials. The commercials imply that not having network coverage is something about which travelers have to be warned. It is expected that travelers will change their travel plans accordingly to avoid these scary places. For instance, in one commercial the car rental agent warns prospective clients of the problems they were having all over the map due to 3G[2] dead zones. Yet, the travelers are not worried at all: They have the Verizon Wireless network "traveling with them." In the motel version of the commercial, the following exchange happens between the receptionist and the traveler:

> Receptionist: Only got one room left and I doubt you want it!
>
> Traveler: Why not?
>
> Receptionist: It's a dead zone. Can't get your calls, your precious emails. It's like you don't even exist.
>
> Traveler: I got the Verizon Network.
>
> Verizon Network guy standing outside the motel surrounded by others: You're good!
>
> Receptionist: Oh! (hands over keys to traveler). The towels are kind of scratchy….

Thus, the commercials also imply that having a Verizon phone makes travel to places possible that are otherwise not accessible.

Several of the iPhone applications currently available deal with the specific traveler angst of needing help in an unfamiliar environment. "Motion Life Travel" promises *live support wherever you go.* The "Emergency Car Care Guide" suggests that we would not want to leave home without it. Similarly, the "Travel Safety Guide" application diligently warns us that without it, one is doomed to fall victim to all sorts of danger:

> It's a dangerous world out there. In today's world, traveling to unfamiliar destinations has become more and more dangerous, for both your personal safety and your pocketbook. It's vital to arm yourself against those who want your money, your person, and even your identity. If you do not want to fall prey to any of the predator, Foreign Travel Safety Tips is a must.

The web-enabled phone takes over a protective role that has been typically played by tour guides. Thus, individual travel outside of organized tours can now still provide some of the same protection that was formerly only available within the confines of a tour group. Even individual travel to foreign places no longer involves the same notion of adventure when it is mediated by web-enabled mobile phone use. One has to either leave the phone at home (which probably asks for tremendous will power and the overcoming of resistance by family and friends), or travel to places where it is impossible to use the phone. Traveling completely outside the network, however, is impossible. The technological layer has clearly defined boundaries which can be crossed. Yet, the other layers are ubiquitous. While one can travel outside the network in a technological sense, culturally, socially, emotionally, etc. one can not.

CONCLUSION

Travel is a major catalyst for economic development in peripheral regions. Travelers' ever greater dependence on the Internet and on mobile phones will encourage the expansion of "the network" to places currently outside of the network grid. Civilization is now where the network is accessible. The periphery is where the network is out of reach. As mobile phone companies continue to expand their coverage, truly remote places will become increasingly rare. The impact of such material components of networks is often underestimated. Human practices are intricately networked with material worlds, and socialities occur with and through these networks (Urry, 2003). The technological layer of the network does not operate in isolation from the other layers. With mobile phone service, other developments might follow, including tourists seeking out the unknown while being backed by mobile phones as their life lines, changing the social and cultural environments of the destinations they visit. Urry (2003) further notes that the power of any network increases with extensions in size or reach. Tourism seems to be a major driver for such network expansion.

The desire to travel within the network also leads to an ever greater need for device standardization, adoption of common mobile phone network standards such as GSM, and strategic alliances between mobile phone service providers. This will likely bring about consolidation in the mobile phone business in the long run as travelers will want to use the same provider wherever they go and use their phones to access web-based contents while on the road without having to pay roaming fees. Thus, travel propels globalization of the network, and with it the spread of network culture (Terranova, 2004).

The emergence of new media technologies creates new preconditions for touristic practices and experiences (Jansson, 2006). New mobile phone applications are developed continuously and will increasingly structure tourism experiences, encourage specific tourist behaviors, and mediate or direct the tourist gaze. Especially, location-based services and mapping applications are believed to become significantly more sophisticated and widely adopted by travelers. Another trend with important implications for travel is the development of augmented reality technologies. New generations of smartphones will be able to project maps over the image of the user's surroundings produced by the phone's camera and displayed on the phone's screen (Markoff, 2009). These maps will contain navigational information, interpretive contents and advertising. They will further focus the tourist's gaze onto the screen and away from the actual, much less informative and engaging surroundings. This clearly stresses the embodied dimension of the network. While augmented reality applications are already being developed, one can only speculate what other developments will emerge from the possibilities afforded by the network.

Web access through mobile phones is becoming an amenity like electricity and running water that many individuals would not want to miss, even when traveling. True escapist tourist experiences are trips to places where mobile phone access is not available. While a select few might find this loss of connection to the network refreshing and a relief, most will probably not be able to enjoy it. While such travel was the only way to travel just a couple of years ago, it is constructed as frightening in the network culture. Thus, "roughing it" for the network generation will likely mean traveling without their web-enabled mobile phone.

This chapter tried to argue that the network has different layers and that the technological base of the network influences various dimensions of travel and everyday life experiences. However, this is not to be understood as a simple deterministic argument. As Barney (2004) points out, the same effects do not occur all the time and not for everyone because specific effects depend on the manner in which the technology is used. What is notable is that if and when impacts occur they are profound, reaching down to even the affective level of our being.

NOTES

1. Note the explicit reference to travel in the language of roaming agreements.
2. 3G = third generation standards for mobile networks

REFERENCES

Aronson, S. H. (1977). Bell's electric toy: What's the use? The sociology of early telephone usage. In DeSola Pool, I. (Ed.), *The Social Impact of the Telephone*, pp. 15–39. Cambridge, MA: MIT Press.

Ateljevic, I. & Doorne, S. (2000). Tourism as an escape: Long-term tourists in New Zealand. *Tourism Analysis*, 5, 131–136.

Barney, D. (2004). *The Network Society*. Cambridge, UK: Polity Press.

Bhattacharyya, D. P. (1997). Mediating India: An analysis of a guidebook. *Annals of Tourism Research*, 24(2), 371–389.

Buhalis, D. & Law, R. (2008). Progress in information technology and tourism management: 20 years on and 10 years after the Internet: The state of eTourism research. *Tourism Management*, 29(4), 609–623.

Castells, M. (1996). The rise of the network society. *The Information Age: Economy, Society and Culture, Vol. 1*. Cambridge, MA: Blackwell.

Clough, P. T. (2008). The affective turn: Political economy, biomedia and bodies. *Theory, Culture & Society*, 25(1), 1–22.

eMarketer (2009a). Smartphone satisfaction numbers. Retrieved July 15, 2009 from www.emarketer.com.

eMarketer (2009b). iPhone downloads up. Retrieved March 19, 2009, from www.emarketer.com.

Eriksson, O. (2002) 'Location based destination information for the mobile tourist, in K. Wöber, A. J. Frew, and M. Hitz (Eds.), *Information and Communication Technologies in Tourism 2002*, Vienna, Austria: Springer-Verlag. pp. 255(264.

Foucault, M. (1995). *Discipline and Punish: The Birth of the Prison*, 2nd ed. New York: Random House.

Galloway, A. R. & Thacker, E. (2007). The exploit. *Electronic Mediations, Vol. 21*. Minneapolis, MN: University of Minnesota Press.

Grewal, D. S. (2008). *Network Power: The Social Dynamics of Globalization*. New Haven, CT: Yale University Press.

Grossberg, L, Wartella, E., & Whitney, D. C. (1998). *Media Making: Mass Media in a Popular Culture*. Thousand Oaks, CA: Sage Publications.

Hansen, M. (2006). *New Philosophy for New Media*. Cambridge, MA: MIT Press.

Harvey, D. (1989). *The Condition of Postmodernity*. London: Blackwell.

Information Society Technologies Advisory Group (2001) 'Scenarios for ambient intelligence in 2010,' Accessed online December 20, 2002 at ftp://ftp.cordis.lu/pub/ist/docs/istagscenarios2010.pdf

Jansson, A. (2002). Spatial Phantasmagoria: The Mediatization of Tourism Experience. *European Journal of Communication*, 17(4), 429–443.

Jansson, A. (2006) 'Specialized spaces: Touristic communication in the age of hyper-space-biased-media,' Working Paper no. 137–06, Centre for Cultural Research, University of Aarhus, Accessed online April 1, 2008 at http://www.hum.au.dk/ckulturf/pages/publications/aj/specialized_spaces.html

Jansson, A. (2007) 'A sense of tourism: New media and the dialectic of encapsulation/decapsulation,' *Tourist Studies*, 7 (1): 5–24.

Koshar, R. (1998). 'What ought to be seen': Tourists' guidebooks and national identities in modern Germany and Europe. *Journal of Contemporary History*, 33(3), 323–340.

MacCannell, D. (1999). *The Tourist: A New Theory of the Leisure Class*. Berkeley, CA: University of California Press.

Markoff, J. (2009). The cellphone, navigating our lives. *The New York Times*, February 17, 2009.

McCabe, S. (2002). The Tourist Experience and Everyday Life. In Dann, G. (Ed.), *The Tourist as a Metaphor of the Social World*, pp. 61–75. Wallingford: CAB International.

McGregor, A. (2000). Dynamic texts and tourist gaze: Death, bones and buffalo. *Annals of Tourism Research*, 27(1), 27–50.

McLuhan, M. (1964). *Understanding Media*. London: Routledge.

Myerson, G. (2001). *Heidegger, Habermas and the Mobile Phone*. Cambridge, UK: Icon Books.

Oertel, B., Steinmüller, K., and Kuom, M. (2002) 'Mobile multimedia services for tourism,' in K. Wöber, A. J. Frew, and M. Hitz (Eds.), *Information and Communication Technologies in Tourism 2002*, Vienna, Austria: Springer-Verlag. pp. 265–274.

Paul, M., Okuma, H., Yamamoto, H., Sumita, E., Matsuda, S., Shimizu, T., & Nakamura, S. (2008). Multilingual Mobile-Phone Translation Service for World Travelers. In *Coling 2008: Companion volume—posters and demonstrations*, pp. 165–168. Manchester, August 2008.

Pearce, P. L. (2005). *Tourist Behaviour: Themes and Conceptual Schemes*. Clevedon, UK: Channel View Publications.

Perkins, H. C. & Thorns, D. C. (2001). Gazing or performing?: Reflections on Urry's tourist gaze in the context of contemporary experience in the Antipodes. *International Sociology*, 16(2), 195–204.

Pew Research (2009). Internet typology: The mobile difference. Retrieved June 11, 2009, from http://pewresearch.org/pubs/1162/internet-typology-users-mobile-communication-devices

Poslad, S., Laamanen, H., Malaka, R., Nick, A., Buckle, P., and Zipf, A. (2001) 'CRUMPET: Creation of user-friendly mobile services personalized for tourism, in *Proceedings of 3G 2001*. London: Institution of Electrical Engineers. Retrieved December, 22, 2002, from http://www.eml.villa-bosch.de/english/homes/zipf/3g-crumpet2001.pdf

Rosh White, N., and White, P. B. (2007) 'Home and away: Tourists in a connected world. *Annals of Tourism Research*, 34 (1): 88–104.

Ryan, C. (2002). Motives, behaviours, body and mind. In Ryan, C. (Ed.), *The tourist experience*, 2nd ed., pp. 27–57. London: Continuum.

Schivelbusch, W. (1986). *The Railway Journey*. Berkeley, CA: University of California Press.

Schmidt-Belz, B., Makelainen, M., Nick, A., and Poslad, S. (2002) 'Intelligent brokering of tourism services for mobile users. In K. Wöber, A. J. Frew, and M. Hitz (Eds.), *Information and Communication Technologies in Tourism 2002*, Vienna, Austria: Springer-Verlag. pp. 265–274.

Sharma, A., & Vascellaro, J. E. (2008). Phones will soon tell where you are. *The Wall Street Journal*, March 28, 2008.

Terranova, T. (2004). *Network Culture: Politics for the Information Age*. Ann Arbor, MI: Pluto Press.

Turkle, S. (1995). *Life on the Screen: Identity in the Age of the Internet*. New York: Simon & Schuster Paperbacks.

Urry, J. (2002a). *The Tourist Gaze*, 2nd ed. Thousand Oaks, CA: Sage.

Urry, J. (2002b). Mobility and proximity. *Sociology*, 36, 255–274.

Urry, J. (2003). Social networks, travel and talk. *British Journal of Sociology*, 54(2), 155–175.

Watanabe, T., Okumura, A., Sakai, S., Yamabana, K., Doi, S., & Hanazawa, K. (2000). An Automatic Interpretation System for Travel Conversation. In *ICSLP-2000*, Vol. 4, 444–447.

Werthner, H., & Klein, S. (1999) *Information Technology and Tourism——A Challenging Relationship*. Vienna, Austria: Springer-Verlag.

Wurtzel, A. & Turner, C. (1977). Latent functions of the telephone: What missing the extension means. . In DeSola Pool, I. (Ed.), *The Social Impact of the Telephone*, pp. 246–261. Cambridge, MA: MIT Press.

CHAPTER FOUR

Membership in the Network

Hardware and Software Development for the Main-Stream Consumer

JOY PIERCE

The Digital Divide is not a conversation that ended with the Clinton Administration. Though the Bush Administration dislodged the notion of a Digital Divide—the gap between people who have computer and internet access and those who do not—by promoting a discourse that assumes digital access and inclusion (National Telecommunications and Information Administration, 2004), these issues still exist (Fox, 2005). In keeping with the previous administration, the Obama administration has not reinstated the Technologies Opportunity Program, also known as TOP, however BTOP, a grant initiative to provide broadband in underserved areas, remains in place (NTIA, 2009). There has been a shift from individuals and the quality of their everyday lives to infrastructure and digital development.

The end of Web 2.0 is likely near. Discussions surrounding Web 3.0 suggest more seamless and intelligent tools (Borland, 2007; Shadbolt et. al, 2006) that will integrate more aspects of an individual's personal and professional life. These tools will likely come in the form of hardware and software. Consumer electronics enter the retail market as virtually old products. While new technologies emerge faster, smaller and lighter than their predecessors, there are many people around the globe who have never owned or accessed the electronic products that so-called mainstream consumers take for granted. Membership in a network society is assumed by electronic developers as well as mainstream consumers.

This chapter integrates data from an exploratory research method called Swarm Scholarship, and case studies from earlier participant observation in an effort to interrogate whether hardware and software developers are attempting to find innovative ways to address the needs and wants of a computer illiterate population. Researchers (Fox, 2006; Haythornthwaite, 2001; Pinkett, 2001) suggest that many undereducated, poor and working class adults have an interest in using new technologies in their everyday lives. Yet they are unable to articulate—without a technophilic vocabulary—how emerging hardware and software may better work toward their interests or needs (Pierce, 2006). This means certain segments of the population are not members of the network society by virtue of education and/or income. I argue that the network developers have contributed to digital exclusion.

EMBEDDED EXCLUSION ⟺ VIRTUAL INCLUSION

Swarm Scholarship is a research strategy developed by Julian Kilker and Josh Greenberg. It is effectively an example of post-global networking. Kilker (2009) explains:

> Swarm methodology includes a small group of heterogeneous researchers (about five to twenty people), who have shared interest in a specific research site but approach it from different analytical perspectives. This group collaborates on the coordination, collection and analysis of the data…it has the potential to be broadly influential for documenting and researching contemporary events that are too large and complex for any single researcher to study (434).

The 2007 International Consumer Electronic Show (CES) was a large-scale, ephemeral event that proved to be an ideal research site for Swarm Scholarship. CES was an exhibition in the epitome of digital exclusivity. More than 140,000 people from around the globe descended upon Las Vegas, Nevada to explore the most cutting-edge new technologies that 2,700 exhibitors had to offer. (Dunion & Szabo, 2007). Some attendees paid as much as $999 to attend the specialty "Knowledge Tracks," test the latest games, and hear keynote speeches by Microsoft Corp. Chairman Bill Gates, Dell founder and Chairman Michael Dell, and Robert Iger, the President and CEO of Walt Disney Company. It is important to note that an Exhibits Plus Pass, which allowed access to exhibits, keynotes and select sessions, was free with advance registration; however, the registrant was required to show two forms of identification to prove affiliation with the consumer electronics industry. The show was not open to the public.

The registration guidelines set forth by the Consumer Electronics Association, the organization that produces CES, demonstrate the subjugation formed by investors and exhibitors as institution gatekeepers of new technology. Drawing on Michel Foucault's (1980) work on the body as an operation of controlling power, the exclusion of people who are not engaged directly in buying, selling or reporting on new technology serves to further separate current high-tech users from potential high-tech users. Here, unlike with new media literacy programs of the new millennium, docile bodies (Foucault, 1977) are completely ignored. Techniques of discipline were once revealed through new media literacy programs funded through government initiatives like TOP. For example, a classroom setting that required attendance, timed sessions or assigned seating exemplified how the body was shaped, manipulated and trained as a form of discipline (Pierce, 2006). While such behavior is certainly problematic, but normalized by those in power, there is room for resistance. Marginalized populations had no participatory power at CES because they were not allowed membership. If individuals are left out of the conversation altogether, there is no room for confrontation or resistance. Institutions are not binary constructions of power, but a site for "constant churning [that] creates gaps for those who wish to challenge existent power relations and existent structures." (Louw, 2001, p. 12) Relations of power are interwoven with other relations such as production and politics, which situate them and are conditioned by them. Further, power is always related to knowledge. Had CES been open to the public or provided just a few hundred tickets to small business owners or two-year technical school students, developers may have gained knowledge into how to better serve more consumers.

Jonathan Sterne (2007) asserts that infrastructure is important, but we don't pay enough attention to it. The CES registration guideline is an example of exclusion and power as discourse and practice. Manufacturers, developers and suppliers of consumer technology hardware, software as well as wireless carriers, cable and satellite TV producers and the new technology knowledge base (educators, financial analysts, government leaders and the media) become the authority on digital development, policy, usership and consumerism through self or formal education and practice.

Network membership is embedded in new technology infrastructure (Star and Bowker, 2006). The central point here is that the manifestation and activity of the CES participants and registrants are not instinctive, but rather things that have been invented and/or learned. It is here that I argue the weakest links in the digital revolution—undereducated, poor and working class adults, and racial and ethnic minorities—are not seen as a desirable consumer market, and so are ignored. The largest barrier to the underrepresented population is embedded in infrastructure. Membership, meaning meeting the criteria to register for CES is reserved for those who understand that new technologies are "sunk into, inside of, other structures, social arrangements and technologies" giving certain populations automatic entrée and therefore a "taken-for-grantedness of artifacts and organizational arrangements" in a technophilic community (p. 231).

New Media research points to Cyberculture (Jenkins, 2006; Nelson and Tu, 2001; Turner, 2006), online identity and social networks (Haythornthwaite, 2007; Merchant, 2006; Monge and Contractor, 2003) or the economic and policy implications (Gillespie, 2007; Benkler, 2006) of the internet on individuals, public and private industry. These research trajectories assume membership in the digital network. While such literature is necessary and important, this chapter contributes to an understanding of how poor, working class and undereducated adults chose to embrace or reject the internet, and how their lived experiences converge or conflict with the ways in which digital network gatekeepers inscribe network membership. I spent four years assisting in a program designed to put refurbished computers in as many homes as possible. Realizing that many of the participants had never used a computer, computer training was part of the requirement for owning a computer and receiving one year of free internet access. A spokesperson for the program emphasized the need for PC home ownership in order to "narrow the gap between the digital haves and have nots." (Kline, 1999) My involvement in this program propelled my search to find viable software and hardware options for poor and undereducated consumers.

My pursuit of hardware and software companies that explicitly addressed the needs of an underrepresented population was non-existent at the outset. The brochures and guides were divided into categories: audio, digital imaging, digital television, home networking, mobile, MP3 players and wireless. After a full day of walking through a sea of men dressed in a uniform of button down or pullover shirts and jeans or khakis—all dripping with electronic devices hanging from their belts—I decided to take a new tact the following day. The course of action the second day was to seek out companies that were involved in the development and manufacture of the $100 laptop program (Nystedt, 2007; Huang, 8 February 2007; UNDP, 2006). The $100 laptop program, now more widely known as One Laptop Per Child (OLPC), was founded by Nicholas Negroponte, former director of Massachusetts Institute of Technology's Media Lab. The program's mission is to provide affordable laptop computers to inadequately educated children in developing countries. While my goal was to seek solutions for the Digital Divide domestically, it occurred to me that a well-known, internationally-supported program such as OLPC would likely have a presence at CES.

Taiwan-based Quanta Computers, the largest manufacturer of notebook computers, and the manufacturer for OLPC was not listed as a CES exhibitor. Advanced Micro Devices (AMD), a global leader in providing microprocessors and media solutions for the consumer electronic industry is another major contributor and was listed as having two booths. The first location I visited was on the plaza. It was a temporary structure—a tent if you will—beside the massive

World Series of Video Games tent and behind a larger, but not as massive, Nokia tent. The Microsoft village—which included one tent devoted to Windows Vista; another for all other Microsoft products; three smaller tents; and a dedicated bus stop—dwarfed even the World Series, Nokia and America Online tents, all surrounding AMD. The person in the AMD tent was unable to engage in any conversation about the company's involvement in the OLPC program. I then visited the main AMD exhibit, located in one of the central halls in the Las Vegas Convention Center. The exhibit was the largest in its region. While there were several staff people on hand, no one at this venue could speak to the OLPC partnership.

It is surprising that AMD and other companies involved in making efforts to usher underrepresented populations into this rapidly growing technophilic society were either not present or lacked personnel informed enough to answer a few questions about the project. The underwhelming enthusiasm of discourse concerning underrepresented populations at CES furthers the argument that Digital Divide criticism can no longer fall completely against the state and policymakers or international and grassroots organizations. New technology challenges the diversity and multiplicity of the social, cultural and economic generated by globalization (McCarthy, 2000).

Many of the representatives at exhibits and booths for AMD, Microsoft, and other microprocessing, computer hardware and software development companies cited consumerism as the likely reason for not providing them with information on products that might be of interest to an underrepresented population. One paid marketing representative said that he did not work for the company directly, and could not speak for the company, but his "guess" was that the cost of exhibiting at CES is in the hundreds of thousands, so the idea was to "get the word out" and sell the company's concept to as many retail buyers as possible. He went on to say that he is certain the company doesn't have a problem with poor people or minorities, but is more interested in people who are already connected. I fear this young representative's guess is as good as many of the other paid marketing associates who worked at a number of the booths and exhibits. It is important to note that CES is not a warehouse retail shopping extravaganza, but a show where retail buyers determine what products they will purchase for national or global distribution; it is where developers meet and make initial plans to collaborate with manufacturers; *the* place where new products are unveiled before becoming available to the general public. Exclusivity is key—being the first to see, touch and test a new product is part of the allure of attending CES.

Originally, I considered the possibility that CES is a venue for people who are already "connected" or have the means to consume the latest technologies. However, Nicholas Negroponte delivered a keynote address for the Technology and Emerging Countries program at 2008 International CES. While the education component of the OLPC project is mentioned, the focus was not on how individuals may benefit from owning a computer, but on the potential markets that are available once people in developing nations are more technologically savvy. Gary Shapiro, president and CEO of CEA emphasized, "As the International CES expands it [sic] reach globally and draws government leaders from across the globe, it creates an important forum for discussion on how best to leverage technology to foster long term growth and development in emerging countries." (CEA, 16 October 2007) Shapiro's enthusiasm for markets over individuals reveals why I was unable to find manufacturers, developers and buyers who were interested in engaging in discourse around how to design, develop and sell information technologies that would empower underrepresented individuals in ways that are meaningful in their everyday lives.

Another question that developed from this research was why a social issue like the Digital Divide was not implicit or explicitly discussed in any literature by CEA or any of its exhibitors. There were displays, press releases and information packets by CEA and the Federal Communication Commission related to privacy issues, environmental pollution in production, green issues for recycling and energy consumption. As Dan Schiller suggests in *Digital Capitalism* (2000), there is a difference between a social issue that involves underrepresented individuals and a social issue that involves targeting a viable market. Privacy issues and going "green" are concerns that may affect the mainstream electronic consumer market. The inclusion of a keynote address by Nicolas Negroponte, founder of a program designed to bridge the global gap in the Digital Divide, does not change this notion. In fact, this sort of blind inclusion—a highly publicized talk about poor, undereducated children in a developing nation—does more to improve the US information technology market than assist underrepresented individuals in ways that are meaningful to their everyday lives in whatever corner of the world they live in.

NETWORK LEGACY⟺MAINSTREAM MEMBERSHIP

Booth and exhibit representatives I spoke with were not always polite. Before saying hello, the representatives looked at my badge to identify my status. I am a racial and gender minority. The lack of racial minorities in the field of information technology is not well documented (Noll, 2000; ITAA, 2003); however, there is considerable literature on the lack of women studying and working in and around new information technologies (Cohoon & Aspray, 2006; Cooper & Weaver, 2003; Wajcman, 2000). At times I was ignored. When addressed, I was oftentimes spoken to in a condescending manner. When I inquired with other swarm collaborators about the way in which they were treated, all said they did not experience the kind of resistance I described. My initial thought was perhaps my status as "press" put me at a disadvantage. Were representatives not interested in talking to people who were not buyers? I had noticed several people with press passes getting information and even interviews. There were other swarm collaborators who obtained press badges. The most striking acknowledgement of difference came when I allowed a collaborator to borrow my press pass for the day. Her experience was markedly different from mine (Paulsen, 2009). In another case, a colleague mentioned being bombarded with information at a tent where I had to go back to the tent entrance and ask for press information. When I initially entered the premises, the person standing at the door looked at me and my press badge then smiled. My colleagues are white; four of the thirteen who attended CES over the course of the pilot study are women.

Once again, I realized I needed to change my research strategy. Turning at once to the knowledge-power dynamic that is often rejected when working with a marginalized population (Sandoval, 2000; Filmer, et al., 2006; Pierce, 2006, 2004); I used my title and university business card in an effort to open a channel of communication with exhibit representatives. It worked. Philippe Bourgois (2002) argues that academic scholars have a history of eclipsing the dominant power structures. Self-awareness of one's mere presence—body language and lived experience—must be taken into account when doing ethnography. Only after introducing myself as "doctor" and presenting exhibit and booth representatives with my business card was I was able to become part of the consumer electronics conversation. I was able to inquire about products and ask digital access related questions by drawing on my own social and cultural capital. My experiences on the second day indicated that status outweighed race and gender. As Allen Johnson (2006) points out, "ignoring privilege keeps us in a state of unreality by promoting the illusion that difference by itself is the problem." (p. 25) My credentials, which were necessary for everyone to gain admission to CES, had become for me a mandatory badge in order to show proof of membership as a technophile through work and formal education. By the end of the second day, I began getting information that could move my research project forward.

Among the seven keynote addresses, eight SuperSessions, numerous press conferences and countless press packets in paper, disc and digital form, only two documents made mention of a technology that would allow people global internet access without the use of a computer. The technology is Voice over Internet Protocol (VoIP). The Federal Communications Commission (FCC) booth supplied a consumer fact sheet that explained that VoIP allows users to make telephone calls using an internet connection. Consumers may use a cordless phone and wire connection or a mobile phone with broadband internet connection. The mandatory requirements for VoIP are a VoIP-specific phone and an internet connection. A user may choose dial-up or broadband; cable modem or wireless internet connection. The fact sheet mentions, "Depending on the VoIP service you purchase, you may need a computer, a special VoIP telephone or a regular telephone with an adapter." (FCC, 2007)

The second press release (Hagan & Hawes, 8 January 2007), from Netgear, a provider of networking products that include wireless routers, media receivers, telephones and other networking solutions, stated that their award-winning dual-mode cordless phone supports landline and Skype internet calls without the need to connect to a computer. The phone retails for $199. David Henry, Netgear's director of consumer products held center stage with representatives from Phillips and IPEVO, a subsidiary of PCOnline out of Taiwan. The three gentlemen were joined at a Skype press conference to introduce VoIP phones. Eric Lagier, Skype's head of business development for mobile devices announced at the elaborately catered conference that Skype was on a course to connect more people to their loved ones for longer periods of time. "We are reaching out to people who do not have a PC," Lagier said. Henry added that connect doesn't mean just computers. "We mean connect to people."

The second half of the press conference was devoted to synergy. Don Albert, general manager of Skype North America explained how a three year old company could have 136 million registered users. "It [Skype] reshapes and impacts lives," he said. The discussion led to the innovative ways Skype finds new revenue schemes for retailers to connect to consumers. The immediate goal is to find ways to connect users with Ebay and PayPal. Ultimately, they envision a "Skype store in a window" that would allow users to click on a tab for services offered by Skype and its partners. Collaboration with Ebay, Google and Yahoo for advertising dollars would ultimately lead to a sort of yellow pages where users could click on a phone to call merchants directly. With each new product and service, a software and likely hardware upgrade may be required. I wondered, how does one upgrade services without a computer?

In an effort to learn more about how VoIP was possible without a computer, I interviewed Prasad Shoroff, product line manager for Netgear. Shoroff explained all that was needed to use VoIP without a computer was the right phone. The Netgear SPH200D phone was one option. The other option was a Philips VOIP841, which would be released February 2007. The Philips phone was not available for viewing. According to Lagier, their model was given to President George W. Bush. The Philips phone would retail for $140. As the exhibit began to fill, I asked Shoroff how one upgrades Skype software without a computer. This was the second time I had asked the question. The first time, Shoroff excused himself and then returned to show me another VoIP phone that would work without the use of a computer. When I posed the question the second time, Shoroff looked at me, tilted his head and replied, "you just hook it up to your laptop."

Mr. Shoroff's response reminded me of conversations with participants in a grassroots initiative designed to teach underrepresented populations how to use a computer (Pierce, 2006). The participants were given a refurbished computer after at least eight hours of instruction. On the day participants were given certificates for computer pick up, they were advised to purchase a power strip. More than 80 percent of the participants I interviewed were flabbergasted to learn they would need to purchase something before being able to use their computers. At the time I did not think their needing to spend ten US dollars was much to ask, but later learned the insult was in assuming they had the money to spare at that particular time.

I visited Skype's website to learn more details about phones that do not require a computer. The brief description for the Philips VOIP841 and Netgear's SPH200D states that it "works without a PC," (Skype, 2007) however, upon closer inspection, the detailed description states that the user may use Skype without having the computer on. In other words, the phones come equipped with software that can be used without the aid of a computer, but it is assumed that the user owns a computer. All software downloads are listed according to computer operating systems. There is no option for downloading directly to the two phones featured as usable without a computer. The only possibility for downloading software upgrades without a computer is through a Smartphone that has a Windows operating system. However, Skype supports a limited number of cellular phone devices, all of which are in Europe. If a user in North America needs to upgrade software, a computer is required.

Vonage, a provider of broadband telephone services through VoIP since 2001, did not make public claims of providing VoIP without the use of a computer. When asked about such prospects, a marketing representative at the Vonage exhibit explained that they service a more mainstream market. When Jeffrey Citron, Chairman and Chief Strategist of Vonage Holdings Corp. speaks of giving customers the flexibility and mobility they want, he does so with mainstream consumers in mind. He said, "The size and scope of the potential VoIP market is vast, with mainstream consumers representing a significant opportunity." (Helies, 8 January 2007) The sentiment was echoed by Vonage CEO Mike Snyder when he announced the next generation of Vonage devices. "As more and more mainstream computer users are becoming subscribers, we want to ensure that our hardware is clever, simple to use and aesthetically pleasing that our customers are proud to display them as part of their home or office." (Yocca, 8 January 2007)

Mainstreamers are people who can create documents using widely distributed computer software programs such as Microsoft, Lotus and Adobe products. The population is also internet savvy; they use online banking and make purchases online. Mainstream computer users are computer users. Even when a discussion of non computer users takes place, as was the case with Skype, the assumption is the consumer is *choosing* at that particular time not to use a computer, not that one does not exist in the home.

Participants I worked with were excited about having a computer in the home. They saw having a computer in the home as an economic and educational equalizer. One thing they had not considered is the need for social capital. More than once a participant confessed that he or she did not know of anyone to email. This is not unlike Payton's findings in "Rethinking the Digital Divide" (2003). She conducted focus groups with ten African-American high school students who planned to attend a four-year college. They said computer access was not a problem, but were aware of the limited number of mentors and role models of color in information technologies (p. 90). The lack of minorities in the network membership not only squashes job possibilities for today's adults (Rodino-Colocino, 2006; Williams, 2005), but potentially for future generations as well.

ARRESTED VIRTUAL DEVELOPMENT

During an evening gathering of information exchange, a swarm colleague informed me that there was a small booth at the end of the exhibit hall at the Sands Expo that might interest me. The next day I found Pepper Computer. The small, hotdog stand-sized booth stood at the end of a row. Had it not been for some direction the evening before, I would have likely not discovered the booth, as it seemed merely an obstacle between a wall which on the other side led to an open gathering spot and the bordering of two aisles, one leading to the massive international gateway and the other to the manic Robotics Tech Zone. The Pepper Computer booth displayed its latest product, the Pepper Pad 3. The handheld device measures 11.4 inches high by 9 inches deep, and weighs 2.2 pounds. It is slightly larger than a VHS tape. The keypad is divided so that one half of the letters on a standard keyboard is on one side of the seven inch screen, and the other half is on the other side. A stylus, which is used on the touch screen, snaps into a well on the bottom of the device. A built-in camera, speakers and microphone allow the user to enjoy multimedia features. There is also a five-way directional pad on the left side of the screen; a scroll wheel is on the right. When asked why the need for a handheld web computer, media relations manager Lindsey Groepper explained that the concept behind the product was simplicity. She said the idea was to create software that anyone would find easy to use. No clicking or going through three steps for one simple task.

"We couldn't find hardware to match the software," she said, so they had a Korean-based manufacturer design the hardware. "They own the design license," she added. The product uses a Pepper Linux operating system. A user is able to carry out a number of online tasks, however, the keyboard configuration is not ideally designed for everyday word processing use. In addition, the product has a usb port but does not offer a DVD option. The Pepper Pad 3 retails for $699 and is available online and through a few retail outlets.

A brochure listed comparable products which included several portable micro personal computers, many of which were less expensive than the Pepper Pad 3. The Hewlett Packard nx6325 notebook computer, an estimated $20 cheaper than the Pepper Pad 3 was cited as more difficult to configure, larger and heavier, and lacking a touch screen and camera. The benefits of purchasing a laptop, according to the brochure were storage space (40 gigabytes versus 20 gigabytes), larger screen, and CD/DVD. Given the options, the Pepper Pad 3 seems useful to people who have access to a desktop or laptop computer at home and/or work.

Pepper Pad 3 was not explicitly designed with an underrepresented population in mind; however its concept fit many of the needs of a person who is not a mainstream computer user. For people who need a user friendly way to access information online, the product is a viable option. However, the price is prohibitive. The design concept of building a computer with simpler software has not fared well in the consumer electronic market. The company's CEO, Len Kawell recently posted to the website's online forum (22 October 2007). He responded to consumers who suggested the company was dead due to lack of response to consumer questions and concerns. He assured posters that the company was not dead. "As you probably can surmise, the original Pepper Pad and Hanbit's Pepper Pad 3 have not sold as well as, say, the iPod," he wrote.

I argue that competition from companies catering to mainstream computer users may be cause for lack of sales for the Pepper Pad 3. People who could use the product cannot afford it, and those who can afford it may choose to spend $699 on a product that is markedly different from a personal computer, such as an elaborate video game console, smartphone, wireless tablet or reading device.

CONCLUSION

Leading digital divide/social media scholars (DiMaggio et al., 2004; Hargittai, 2004; and Kvasny, 2006; Selwyn, 2004) have moved from the binary have/have not discussion to a more nuanced argument of digital inclusion and effects. Wakeford argued in 2004 that a researcher would be "hard pressed to find any body of work which worked through the epistemological and methodological implications of the multiplicities, incongruities and partialities of knowledge outside mainstream social groups and cultural locations,"(130) despite its central importance in understanding certain social and cultural contexts. Yet there remains little literature that explicitly articulates the systematic ways in which underrepresented populations continue to be excluded from membership despite grassroots efforts and government initiatives. Such exclusions and oversights will likely lead to further economic, social and educational disadvantages that have historically plagued underrepresented populations living in a post-global society. This chapter begins to address Wakeford's call to action while extending the use and effects discussion to include emerging technology researchers and developers.

REFERENCES

Allen, B. (2002). Goals for emancipatory communication research. In M. Houston & O. Davis (Eds.), *Centering Ourselves: African American Feminist and Womanist Studies of Discourse.* Cresskill, NJ: Hampton Press, Inc.

Benkler, Y. (2006). *The Wealth of Networks: How Social Production Transforms Markets and Freedom.* New Haven, CT: Yale University Press.

Borland, J. (March/April 2007). A smarter web: New technologies will make online search more intelligent—and may even lead to "web 3.0." *Technology Review,* 110 (2), 64–71.

Bourgois, P. (2002). Ethnography's troubles and the reproduction of academic habitus. *International Journal of Qualitative Studies in Education, 15*(4), 417–420.

CEA. (8 January 2007). History—40th anniversary of the International CES: Arlington, VA: Consumer Electronics Association.

CEA. (16 October 2007). New CES Program Explores Role of Technology In Emerging Economies. Arlington, VA: Consumer Electronics Association.

Cohoon, J.M. & Aspray, W. (2006). *Women and Information Technology: Research on Underrepresentation.* Cambridge, MA: MIT Press.

Collins, P. H. (2000). *Black Feminist Thought: Knowledge, Consciousness and the Politics of Empowerment* (2nd ed.). New York: Routledge.

Cooper, J. & Weaver, K. D. (2003). *Gender and Computers: Understanding the Digital Divide.* Mahwah, NJ: Lawrence Erlbaum Associates.

DiMaggio, P., Hargittai, E., & Celeste, C. (2004). From unequal access to differentiated use: A literature review and agenda for research on digital inequality. In K. Neckerman (Ed.), *Social Inequality.* New York: Russell Sage Foundation.

Dunion, T. & Szabo, S. (2007). 2007 International CES kicks off 40th anniversary with global array of the hottest C.E. products from 2700 exhibitors. International Consumer Electronics Show. Press Release, Las Vegas, NV: Consumer Electronics Association.

Ebron, P. & Tsing, A. (1996) In dialogue: Reading across minority discourses. In R. Behar & D. Gordon (Eds.), *Women Writing Culture.* Berkeley: University of California Press.

FCC. (2007). FCC consumer facts: Voice over Internet protocol. Washington, DC: Federal Communications Commission.

Filmer, A., Pierce, J. & Dolan, K. (2006). Confronting difference as qualitative researchers: Authoethnography as a site for intervention, connection, action. Panel presented at the

National Communication Association Annual Meeting, San Antonio, TX.

Foley, D. & Valenzuela, A. (2005). Critical Ethnography: The politics of collaboration. N. Denzin and Y. Lincoln (Eds.), *The Sage Handbook of Qualitative Research* (3rd ed., 217–234). Thousand Oaks, CA: Sage Publishing.

Fox, S. (2005). Digital divisions. Washington, DC: Pew internet & American life Project. Retrieved 2007 October 1 from http://www.pewinternet.org/pdfs/PIP_Digital_Divisions_Oct_5_2005.pdf.

Gillespie, T. (2007). *Wired Shut: Copyright and the Shape of Digital Culture*. Cambridge, MA: MIT Press.

Hagan, D. & Hawes, L. (8 January 2007). Skype calling advances: Netgear.

Haythornthwaite, C. (2007). Social networks and online community. In A. Joinson, K. McKenna, U. Reips & T. Postmes (Eds.), *Oxford Handbook of Internet Psychology* (121–136). Oxford, UK: Oxford University Press.

Helies, M. (8 January 2007). Beyond broadband voice: Vonage and Earthlink team to offer Wi-Fi access. Holmdel, NJ: Vonage.

Huang, R., Tzeng, D. & Shen, S. (2007). Quanta poised for profits of OLPC shipments reach 100 million. Retrieved 1 October 2007, from http://www.digitimes.com/systems/a20070208PD212.html

ITAA. (2003 May 5). Report of the ITAA blue ribbon panel on IT diversity, National IT Workforce Convocation. Arlington, VA: Information Technology Association of America. Retrieved 2007 October 4 from http://www.itaa.org/workforce/docs/03divreport.pdf

Jenkins, H. (2006). *Fans, Bloggers, and Gamers: Exploring Participatory Culture*. NY: New York University Press.

Kilker, J. (2009). Exploring a new methodology: background, planning, and lessons from the 2007 tradeshow 'swarm' project. *Social Identities: Journal for the Study of Race, Nation and Culture, 15*(4), 433–446.

Klein, N. (2002). *No Logo*. NY: Picador Press.

Kvasny, L. (2006). Cultural (re)production of digital inequality in a US community technology initiative. *Information, Communication & Society, 9*(2), 160–181.

Ladson-Billings, G. (1998). Just what is critical race theory and what's it doing in a nice field like education? *Qualitative Studies in Education, 11*(1), 7–24.

Lorde, A. (1984). *Sister Outsider*. Freedom, CA: Crossing Press.

Mansell, Robin (2004) Political economy, power and new media. *New Media & Society, 6* (1). pp. 74–83.

McCarthy, C. D., G. (2000). Governmentality and the sociology of education: Media, educational policy and the politics of resentment. *British Journal of Sociology of Education, 21*(2), 169–185.

National Telecommunications and Information Administration (2004). A nation online: Entering the broadband age. In U. S. D. o. Commerce (Ed.).

National Telecommunications and Information Applications. (2009). Office of Telecommunications and Information Administration homepage. Retrieved 2009 June 18 from http://www.ntia.doc.gov/otiahome/otiahome.html.

Nelson, A., Tu. T. & Hines A. (2001). Introduction: Hidden circuits. In A. Nelson, T. Tu & A. Hines (Eds.), *Technicolor: Race, technology and everyday life*. New York: NYU Press.

Noll, K. (2000). Grants to train US workers for high-tech jobs often filled by foreign workers. Retrieved 2007 November 3, from http://www.doleta.gov/sga/awards/00–104award.cfm.

Nystedt, D. (2007). One million OLPC laptop orders confirmed. Retrieved 1 October 2007, from http://www.networkworld.com/news/2007/021507-one-million-olpc-laptop-orders.html.

OLPC. (2007). One Laptop Per Child: Mission. Retrieved 6 November 2007, from http://laptop.org/en/vision/index.shtml.

Payton, F. (2003). Rethinking the digital divide. *Communications of the ACM, 46*(6), 89–91.

Paulsen, K (2009). Ethnography of the ephemeral: Studying temporary scenes through individual and collective approaches. *Social Identities: Journal for the Study of Race, Nation and Culture, 15*(4), 509–524.

Pierce, J. (2004). Teaching the internet to the "other" half. *Television and New Media, 5*(2), 141–146.

Pierce, J. (2006). *Communication Unplugged: A qualitative analysis of the digital divide.* Unpublished Doctoral Dissertation, University of Illinois at Urbana-Champaign.

Pinkett, R. (2001b). Community technology and community building: Early results from the Camfield Estates-MIT creating community connections project. Paper presented at the 43rd Annual conference of the Association of Collegiate Schools of Planning, Cleveland, OH.

Rodino-Colocino, M. Laboring under the digital divide. *New Media and Society, 8*(3), 487–511.

Sandoval, C. (2000). *Methodology of the Oppressed*: University of Minnesota Press.

Schiller, D. (2000). *Digital Capitalism: Networking the Global Market System.* Cambridge, MA: MIT Press.

Selwyn, N. (2004). Reconsidering political and popular understandings of the digital divide. *New Media and Society, 6*(3), 341–362.

Shadbolt, N. Hall, W. & Berners-Lee, T. (2006). The semantic web revisited. *IEEE Intelligent Systems.* IIEE Computer Society. Retrieved 21 June 2009 from http://eprints.ecs.soton.ac.uk/12614/1/Semantic_Web_Revisted.pdf.

Skype. (3 February 2007). Cordless Phones, from http://us.accessories.skype.com/DRHM/servlet/ControllerServlet?Action=DisplayCategoryProductListPage&SiteID=skype&Locale=en_US&Env=BASE&parentCategoryID=4141800&categoryID=7032500

Star, S. & Bowker, G. (2006). How to infrastructure. In L. Lievrouw & S. Livingstone (Eds.), *The Handbook of New Media.* (2nd ed., 230–245). London: Sage Publishing.

Sterne, J. (2007, 29 September). Synthesis and discussion. Paper presented at the Frontiers in New Media Symposium, Salt Lake City, UT.

Turner, F. (2006). *From Counterculture to Cyberculture: Stewart Brand, the Whole Earth Network and the Rise of Digital Utopianism.* Chicago: University of Chicago Press.

UNDP. (2006). $100 Laptop Project Moves Closer to Narrowing Digital Divide. Retrieved 1 October 2007, from http://www.digitaldivide.net/articles/view.php?ArticleID=581

Wajcman, J. (2000). Reflections on gender and technology studies: In what state is the art? *Social Studies of Science, 30*(3), 447–464.

Wakeford, N. (2004). Pushing the boundairies of new media studies. *New Media and Society 6*(1), 130–136.

Williams, K. (2005). *Social networks, social capital, and the use of information communications technology in socially excluded communities: a study of community groups in Manchester, England.* Unpublished Doctoral Dissertation, University of Michigan.

Yocca, J. (8 January 2007). Next generation Vonage devices to get sleek new look: Vonage partners with frog design to create new product design. Holmdel, NJ: Vonage.

CHAPTER FIVE

Voicing and Placement in Online Networks

RADHIKA GAJJALA & ANCA BIRZESCU

INTRODUCTION

Scholars in Cyberculture studies/Internet research continue to grapple with issues of race, ethnicity, representation, identity, development and globalization through various critical and not-so critical lenses. As such scholars' work indicates, it is important to consider the dual nature of the Internet. On the one hand, the Internet represents a site where rich and sustained interactions are constitutive of cultures and, moreover, a culture in itself. On the other hand, the Internet is also a cultural artifact produced by people having contextually situated goals and interests (Hine, 2000). It is therefore a particular domain of material culture which, on the one hand, provides people with the space for enacting core values, practices and identities and, on the other hand, can be molded by people to specific purposes (Miller and Slater, 2000). Critical feminist scholars (e.g., Enteen, 2006; Gajjala & Altman, 2006; Kolko, Nakamura & Rodman, 2000; Nakamura, 2002) therefore underscore the importance of intricate everyday practices and socio-cultural hierarchies. They point to how our daily practices contribute to the shaping of online identities, hierarchies, and processes of inclusion and exclusion (gendered, raced, classed online subjectivities). Likewise, attention should be paid to the effects of cyberspacial praxis in the offline world. Thus, for instance, the phenomenon of outsourcing due to the employment of foreign workers as well as the increase in philanthropy online, produces effects on how global social networks online are shaped.

As Miller and Slater (2000) explain, freedom online, like off-line, is always constructed, and we should be cautious when asserting that the Internet provides liberated and fluid identities. For instance, the concept of cybertyping is employed in the examination of how race is signified in the discursive and rhetorical space represented by the Internet (Nakamura, 2002). Nakamura uses the concept to describe "the distinctive ways that the Internet propagates, disseminates, and commodifies images of race and racism." (p. 3) More specifically, cybertyping is the process where "computer/human interfaces, the dynamics and economics of access, and the means by which users are able to express themselves online interact with the…ideologies regarding race that they bring with them into cyberspace." (p. 3) The seemingly progressive nature of the Internet is shown to be contrasted by the practice of identity tourism on the Internet, which only reiterates gender and racial stereotypes and, even worse, evidences the consumption and commodification of racial difference. The visions of 'postracial democracy' apparent in the discourse surrounding the Internet, are more than often a device working toward a cosmetic cosmopolitanism or multiculturalism that seeks to mask the reality of racism (Nakamura, 2002).

Thus the issues of representation of race and ethnicity in cyberspace as well as of material and cultural access to Information Communication Technologies (ICTs) are significant areas of research that connect cyberculture studies research to communication studies scholarship. Cyberculture researchers and Communication researchers are also very interested in the study of "Social network systems" online such as Facebook, MySpace, Friendster and so on, especially in regards to youth cultures and fan cultures (see work by scholars such as Ellison and Boyd among others).

In the current book chapter we draw on critical lenses that were formed through the study of race, ethnicity, development and socio-economic globalization in cyberspace and examine some issues concerning subjectivity, voice and agency in social network systems online. In particular we focus on the issue of voice as it emerges in social networks in order to explore the implications of such voicings in a global economy with a global labor force.

Of late, cyberspace has been a site for marginalized voices to emerge in activism and protest as diverse populations gain access to the Internet and related literacies. However, rather than stop at a euphoric celebration of this emergence of voice from thus far marginalized groups we try to investigate what implications these emerging voicings might have for the existing and emerging structures of power. We wish to understand how voice emerges in any of the contexts we examine in order to understand where oppression shifts to when particular marginalized groups gain voice within structures of globalization. Globalization processes include material and discursive hegemonies produced at the intersection of the economic, the cultural, and the social, and are mediated in multiple ways through old and new mediascapes. These processes feed into economic and cultural local formations. Global technospaces are produced through and are a consequence of economic globalization.

We use the term "voicings" to suggest the shifting nature of voice as acts of speaking occur within various contexts. We signal how "voice" is a construct based in situated, contingent speech acts shaped through existing power hierarchies. In such venues, the marginalized speaker emerges as a speaking agent as s/he speaks through the cracks and fissures—ruptures—that occasionally or accidentally permit subaltern speech to be heard in the mainstream. After voice emerges in this manner and finds a way to carve a space for the agent, the speaker is subjected/disciplined into the existing power structure. Thus, for instance, the Dalits (members of a historically ill-treated lower caste in Indian society) of India have voice online and in various activist venues offline and online. However, they are heard only when they voice their issues and concerns within frameworks recognizable within existing discursive logics shaped by current power hierarchies.

What follows is an attempt to understand the workings of discourse and practice in cyberspace that continue to produce the subaltern "non-technical" in ways that further serve to disempower such a subaltern identity within global spaces of speech and action. Therefore, we examine specific examples of how subalternity and subaltern causes are re-produced through events and representations in online spaces.

Part One

DEFINITIONS AND THEORY

"Voice," marginalized or mainstream, does not emerge from the sheer need for the existence of a subject position. Rather, the needs and pains of the unspoken of and unheard of populations the world over get voiced within contexts that make it possible for them to emerge. Thus, as Spivak has suggested through her well-known work "Can the subaltern speak," speech acts performed in resistance get rewritten, read and voiced as subject positions that are slottable within an emerging mainstream logic. The making visible of the past and present practices of erasure of particular subject positions leads to the struggle for legitimization. This in turn leads to the formation of a subject position in future generations for which there is an acceptance (even if continually troubled and negotiated) within the mainstream. This acceptance, no doubt, is riddled with contradictions and nuances. The contradictions and nuances based in such a mainstreaming of a thus far subaltern identity leads to a slicing of the group in ways that reclassify hierarchies—often along global class lines. These global class lines are often only partially visible because of the slippages in cultural, racial, ethnic, gendered and national identities that produce contradictions in voicings.

As we have seen in the discussion above, there has been considerable celebration over how the Internet allows various marginalized voices to access the global and how empowering that is. Simultaneously as there is discourse about the Internet in this manner, we see that there are many non-governmental organizations (NGOs) springing up all over the world to champion some cause or the other, running the gamut of human issues from environmental protection to human rights to development assistance. NGOs can be, according to a classification of the World Bank, operational (involved mainly in designing and implementing projects), or advocacy NGOs, that defend and promote a given cause; nevertheless, there are NGOs that can fulfill both functions at the same time.

Among development NGOs representing the causes of underrepresented or marginalized populations, there is a particular type, namely, the Grassroots Support Organization, which functions as an intermediary, networking institution. It fosters links between the beneficiaries (local, disadvantaged rural or urban groups and individuals) and the otherwise inaccessible donors, financial institutions, etc. This is the type of NGO that Kiva.org, the object of the current critical scrutiny, represents. Kiva is the sort of nonprofit that provides services indirectly to other organizations that support the poor or perform coordinating or networking functions." (Carrol, 1992) Kiva.org makes use of the Web 2.0 tools precisely to fulfill its networking function. Thus, in our view, Kiva.org stands for a Grassroots Support Organization. Such Development NGOs have increasingly found their way to the Internet, and NGOs online have been key actors in bringing forth various economically marginalized voices into the global economic mainstream.

These NGOs have developed various persuasion formats and tactics to engage the mainstream, as Clifford Bob (2005) observes in his work on "Marketing Rebellion." In employing such tactics and formats, the NGO's desire to give voice to marginalized populations and particular causes and to connect them to the global. Social networking technologies are also being used in these efforts. Kiva.org, for instance, is considered to be a very user-friendly platform for microfinance. In this paper, we draw on a continuing study of Kiva.org that is focused around economic development activities where borrowers and lenders are linked with each other in online space. As described in its "About" section, "Kiva is the world's first person-to-person micro-lending website, empowering individuals to lend directly to unique entrepreneurs around the globe." (http://www.kiva.org/about) The interplay of online and offline connections and activities associated with this site impacts who speaks and how they speak.

Critical scholars focusing on voice have previously asked questions such as "who speaks?" (Alcoff, 1993) and examined voice and silence relation to material and discursive hierarchies, while liberal cyberfeminists and technophilic celebrators of the Internet make claims of the Internet's ability to allow voices from the margins to speak up (Landow, 1992, Rheingold, 1991). However, an examination of multiple and even chaotic voices as they emerge in these social networks allows us to highlight a different kind of disciplining role played by the technical interface and design. At the same time as the individual begins to feel empowered by the ability to speak up and back in such networks—there is a quick and simultaneous appropriation occurring that swiftly *places* this voice into a slottable position that can be located and categorized as known and knowable. However, what is crucial to note here is that, while appropriation occurs, there is also a certain kind of empowerment also occurring simultaneously.

This apparent contradiction has been noted in pedagogical settings. Feminist teachers have implemented computer-mediated communication in their teaching to see if it allows women and girls voice online. In such attempts, discursive context and discursive agency are brought to bear on the writing process so that students write for self-transformation and social change. However, contrary to some understandings of critical and feminist pedagogies, "giving voice" to students or empowering the student voice to talk back to power is not enough. Power is a complex interplay of context, cultural formation and individual action that gets produced and reproduced in interaction. They observe, in writing of how computer-mediated environments in the classroom can play out, that:

> Viewing the contested terrain of computer-mediated education within the classroom environment, one discovers a quizzical corporeal rhetoric etched upon and woven through the body of the computer-augmented instructor, a performance that emerges through the interplay of place and space. In other words, the place of instruction—a site of walls, desks, boards, and rulers—evokes strategies of social order, calling forth a space of interaction—utterances, pauses, statements, and silences—tactics of resistance that often demean both instructor and student. Here, one confronts a far more complicated arena of human discourse than that often celebrated in public promises about the wired classroom. Indeed, one discerns the edges of a new level of struggle: efforts to teach students the practices through which their "lessons" are constructed bump against the soft wall of unspoken desire of many teachers who would ensure that these "empowered" students do not gain the tools to displace instructors altogether (Wood & Fasset, 2003, p. 294).

Similarly, in social networks—whether the explicitly leisure oriented ones such as Facebook and MySpace or those formed to empower and finance individual groups from marginalized communities such as Kiva—power plays out in a nuanced, complex way: It allows an emerging global capitalist status quo to reassert itself while allowing/making possible for particular formations of marginalized voices to be centered. Thus, the way in which power plays out shifts to work around these centered marginalized voices by making them seem the exception to the rule, by assigning to them some values associated with the existing mainstream or by exoticizing them.

Does this mean that the empowerment and voice gained by marginalized groups in such a setting—whether momentary or not—should be discounted as insignificant? No, that is not what we are arguing. Instead we signal the process through which the emergence of multiple marginalized voices at social network interfaces becomes possible through contradictions. At the same time we point to the simultaneous re-placement of voice and identity in an emerging power structure. This in turn urges us to examine such shifts if we are to understand what the paradigm and technical interface of "network society" might mean for us socially, culturally and economically. We are also interested in questioning whether just the fact of emerging voice in and of itself is enough for empowerment of marginalized populations. In addition, we note that the oppression and silence shifts as hierarchies subtly slide into a re-organization that is apparently "inclusive." Thus, what expected and unexpected forms might this empowerment take and what paths, trajectories and networks emerge? Also, what does this moment of empowerment mean for the various actors and contexts within and through which the voice emerges? Where does privilege and hierarchy shift, and who and what does it oppress and why? Online networks—built through technologies that foster quick interactive moments through instant messaging, texting, frequent updating of status messages, continual blogging and updating—allow for instant communication that is conducive to subversive and evasive one-on-one contact while simultaneously public and open to surveillance. Thus, the way power plays out in such contexts is nuanced and complex. A group of teenagers may produce "egocentric networks" on Facebook and MySpace as suggested by Boyd and Ellison (2007). But these same spaces also function panopticon-like to discipline subjectivities through uni-formed practices of work and play within such networks.

Further, what is happening in these online settings signals what Kamari Maxine Clarke (2004) notes in her work on "Yoruba Networks" as a shift from "spatial structure and organization to the process of producing space and identity with new technologies of knowledge." (p. 24) In such online networks, there is an interplay of multiple locals/locales at the site of the internet-mediated global. These in turn become global and resituated and placed in various other different locals. This leads to a production of uniform social practice and norms through the use of the "third space" (Soja, 1996) of online networks. Online networks can be viewed in this sense as allowing for the formulation of "thirdspace." Here, third space represents—in Soja's reconceptualization of human interaction around the concept of space—"a product of a 'thirding' of the spatial imagination, the creation of another mode of thinking about space that draws upon material and mental spaces of the traditional dualism but extends well beyond them in scope, substance and meaning." (p. 11) This "third space" reinstates particular local hierarchies in different locales worldwide and names them as global. Clarke (2004) writes that

> Shifts in capital, institutions, historical trajectories, and rules of citizenship resulted in the need to rethink not only the ways that agents in deterritorialized reshaped spaces to produce new meanings of place, but, most centrally, the relationship between particular places and…institutional "networks" that connect them. This shift from the local to the interplay between local and the global…is critical for the understanding of how zones of interaction are not only imagined and shaped but also aligned along domains of knowledge and power." (p. 25)

While Kamari Maxine Clark is writing about deterritorialized space and territorialized subjects in offline networks, what she observes is relevant in online network contexts as well.

NETWORKS, SOCIAL NETWORK SITES AND NETWORKING ON KIVA.ORG

Kamari Maxine Clarke's (2004) conceptualization of networks in terms of territorialized subjects and deterritorialized spaces and Tiziana Terranova's (2004) discussion of network culture—she urges us to "think simultaneously the singular and the multiple, the common and the unique"—are useful to us as we draw from in-depth examinations of social network sites to point to the processes through which these social networks shape voice online. Differentiating between social network sites and the practice of social networking, Boyd and Ellison (2007) lay out what they see as the main features of social network sites:

> We define social network sites as web-based services that allow individuals to (1) construct a public or semi-public profile within a bounded system, (2) articulate a list of other users with whom they share a connection, and (3) view and traverse their list of connections and those made by others within the system. The nature and nomenclature of these connections may vary from site to site (p.)

They state that social *networking* does not necessarily happen on social network sites, for that is not the primary practice on such sites. They contend that social network sites are unique not because they allow meetings between strangers, "but rather they enable users to articulate and make visible their social networks…" therefore, they claim that a critical organizing feature of these social network sites emphasizes an "articulated social network." (Boyd and Ellison, 2007) However, they write about social network sites such as Friendster, MySpace and Facebook. The site we examine in the present paper, Kiva.org, is primarily about social *networking*. Users connect for economic reasons and the connections are mostly between borrowers and lenders for economic transactions focused on development and empowerment.

In the present chapter, we use specific instances of identity production and network interaction in Kiva.org. As a result, we develop the idea of emerging global networks by locating networks at online/offline intersections of imagined space and offline physical, material, socio-cultural and economic placement that includes practices of everyday life. We make a shift away from looking at representation of identities as stereotypes and essentialized identities across time and context. In other words, we make a shift away from what Kamari Maxine Clarke calls the "geography of subjects within self-contained spaces." (p. 2) Such a geography of subjects asks us to "relegate heterogeneity to homogeneity" while at the same time insisting on clear (yet artificial, even if political) boundaries between the named categories. It is this geography of subjects that is mobilized in the service of contained and disciplined multiculturalism. This is not to deny that examination of subjects within self-contained spaces provides important insights into stereotypes and the politics of representation within global media environments. Nor is this meant to discount the body of work that does this in relation to race and ethnicity in cyberspace (which includes one of the co-author's own continuing work). However, in this chapter we make a shift that allows us to look at the ways in which these networks legitimize dominant power hierarchies and how legitimation of membership in fields of power takes place. In order to examine the specific located intersections that produce specific identities and voices within networked environments online, we build on Clarke's conceptualization of "deterritorialized spaces" and "territorialized subjects."

In what follows, we examine ways in which subaltern voice is placed in the socio-economic network Kiva.org in order to get to an understanding of how social networks contain territorialized subjects in seemingly deterritorialized space.

Part Two

KIVA.ORG

Kiva.org represents a relevant case in connection to practices of identity production and network interaction in the context of online social networks. The main objective of this Internet enabled program is to raise debt capital for microfinance institutions in developing countries directly from lenders/social investors, under the condition of "client impact transparency on the Internet." This condition that underpins the intermediary role played by Kiva.org is achieved by Kiva's volunteering fellows through field journaling and interviews with clients. The Internet mediated connection provided by Kiva organization between possible lenders and individual entrepreneurs, both situated at opposite ends of the economic status and in different geopolitical spaces, seems to enhance the microfinance practices. The most important features on Kiva's network site, made possible by the Web 2.0 tools, are the diaries of Kiva volunteers. These record the impact of the microloans on the clients' businesses and lives, the pool of profiles of the clients/borrowers that prospective lenders may access in order to decide which individual/s to lend money to, as well as the self-profiles of lenders or lending teams and groups. The networking practices fostered by Kiva help increase the chances of entrepreneurs in economically disadvantaged locations to get loans by the exposure of their cases to a much larger audience that, without the ICT solutions offered by Kiva, would be restricted to local contexts.

However, in the context of the more general discussion about the global horizontal communication networks fueled by "mass self-communication"—the communication process specific to network society, which presupposes "the multimodal exchange of interactive messages from many to many both synchronous and asynchronous" (Castells, 2007, p. 246)—close textual and visual analysis of Kiva interface shows instances where the Web 2.0 tools, intended to help network people in seemingly global deterritorialized online spaces, reinscribe/reposition the Other within a spatial/territorial conception of the Internet.

Castells's (1996) conceptualization of network society may be viewed as having common grounds with Ellison's explication of "networking" (initiating relationships/ building communal identities). According to him, networks are "open structures, able to expand without limits, integrating new nodes, as long as they are able to communicate within the network, namely as long as they share the same communication codes." (p. 470) Castells (1997) relates the idea of networks to empowering identity building processes, and stresses the fact that the social construction of identity necessarily takes place in a context informed by power relationships. In this context, he defines *project identity* as a particular category which takes shape when "social actors, on the basis of whichever cultural materials are available to them, build a new identity that redefines their positions in society and, by doing so, seek the transformation of overall social structure." (p. 8) An analogy may be thus easily made with Kiva enterprise itself, where, by means of the cultural materials represented by Internet technology, social actors are seeking apparently to change current economic models. Castells asserts that network society calls for new forms of social changes, given the fact that it is based on the dialectical interplay between local and global for an increasing number of individuals and social groups, and because of the separation between power and experience on distinct time-space frames. This leads to the shrinking of civil societies, since the power-making in the global network does not really need the association of individuals in specific cultures and societies. As a result, there is a communal resistance in the network society, which emerges from the attempt to reconstruct defensive identities around communal principles. By the same token, opposing the spatial, territorial conception of the Internet to the network metaphor, Tubella (2006) argues that Internet may play an important part in the reconstruction of a communal self since it entails the replacement of the old idea of territorial-oriented community and belonging with identity aspects such as connectiveness and cooperation. For instance, Giddens (1991) emphasizes the role of technology, through the function of mediating social relations, in the dissociation of ethnicity (ethnic self) from a particular space or period of time.

Nevertheless, along with the opportunity in the network society for counter power offered by the horizontal communication networks, there is the attempt on the part of the powerful to rearticulate new forms of hegemony by power making acts ranging from legislation labeling Internet users as pirates, to purchasing social networking sites in order "to tame their communities." (Castells, 2007, p. 259)

Within the polemics regarding the dominant space and place-oriented conceptions of the Internet, and in contrast to the seemingly positive perspective of a network society, Enteen (2006) contends that hegemonic spatial conceptions of the World Wide Web extend and reiterate the contemporary geopolitical inequities in our offline world, rather than echoing "the connectivity and interdependence suggested by the term 'web'." (p. 233) Territorial metaphors of the Internet have powerful ideological implications in that, like earlier colonial imagery employed to describe territorial incursion, they code social control, power of conquest, and hierarchical positioning into exploration and inhabitation, and thus might forcibly dislocate other discourses describing electronic communication (Enteen, 2006). As Enteen (2006) explains, "rather than recognizing the networks formed through online information exchange, the prevailing images of the internet and world wide web locate individuals, not to mention data, within spatial coordinates." (p. 229) Against an apparent deterritorialized condition made possible by cyberspace, Kaplan (2002) argues likewise that "a closer look reveals location and materiality in the mobility and disembodied discursive practices of new information technologies." (p. 3)

Sometimes the website highlights of personal information in the lenders' profiles or the information in the lending team profiles allude to their understanding of the cooperation/networking enterprise of Kiva.org. Such are the following quotes by an American lending couple in the "About" section of their profile page: "cooperation is the key to making people feel less alone in this world." (Kiva Lender: Charles) Likewise, the info in a lending team's profile reads: "We loan because we care about the suffering of human beings....Those of us who know we are one human family." (Kiva Lending Team) However, the principle of mass self-communication fueling this network site, that would ideally help transcend spatial metaphors and epistemes underlying hierarchical geopolitical sites in the name of globalization, is being hijacked by the ways in which the Web 2.0 networking tools are deployed within the economy of the Kiva.org website.

Kiva fellows' journals and profiles with verbal and visual content about borrowers, both individual and group, is the most critical aspect for Kiva.org strategy to create authority and credibility within the world microfinance system. Not only is this a means to persuade prospective lenders of the viability of person-to-person microloans, to assure the transparency of the impact of the loans on clients' livelihood, and to foster a degree of immediacy between lenders and borrowers, but it also produces specific subjectivities which are informed by the general objectives of the organization. The profiles include biographical information about the borrowers and details about their businesses or business plans so that the overall content makes a strong case that facilitates website viewers' decisions to start lending.

The symbolic functions of the visuals are highly relevant to the degree that the contrast between absence and presence in the discourse produces othered subjectivities characterized by powerlessness, poverty and piteousness. In comparison to lenders' profiles, there are the borrowers' profiles where the fact that the snapshots represent the borrowers looking down or, in any event, not making direct eye contact with the camera is not just a simple coincidence. The photographical discourse refuses thus agency and the possibility of scrutiny to the borrowers, objectifying them and eclipsing their opportunity of "looking back" in a dialogic encounter with the camera. The camera stands here for the eyes of the prospective lenders since the photographic perspective substitutes the lenders' perspective. By contrast, in the lenders' profiles, the snapshots show these people looking directly into the camera eye in an individualized, uncensored process of knowing and scrutinizing.

The subject positions produced by profiles and journals' discourse take shape through a storyline prototype. There are three stages worth noticing in the economy of this storyline, such as the initial situation of the borrower, the narrative pointing to the necessity of the loan, and the post-loan stage emphasizing the positive outcome of the loan. The beginning of the storyline is punctuated by dramatic overtones that render each case unique. Both verbal and visual levels attempt to acknowledge the voices of these entrepreneurs-borrowers and, most importantly, make their presence visible to an outside audience which, already engulfed in a capitalist system fueled by credit and consumption, would have a hard time otherwise fathoming their standpoints.

The profile of the Amelia Ubilluz Mejia's group, a borrowers' group in Peru, displays succinctly three women's life trajectories; the three are in their forties, two of them are married with 3 and 8 children respectively, and they are small shop or street vendors in need of loans to keep on track their small businesses. Most importantly, the readers will easily understand that the improvement of their businesses through micro-loans will implicitly make possible, in the long run, the women's dream stated in the profile, namely, that their children "enter a profession." (Amelia Ubilluz Mejia's Group)

While storylines in the basic profiles delineate subject positions encumbered by unmanaged needs and problems but at the same time determinate and creative in finding venues to overcome challenges and hopeful of a better near future, the journals' narratives make surface renewed subjectivities, in sync with the positive outcomes produced by the loans on their businesses/ lives. Along with voicing borrowers' gratitude to Kiva and Kiva's lenders, the journals' content stands for an ethnographic exercise that familiarizes viewers/readers with representations of local subjects and knowledge. Such is the example of Kiva's fellow field journal about Margaret, a women from Uganda, who used the loan received via Kiva.org for her multiple businesses. The Kiva fellow onsite wondered initially why individuals in the entrepreneurs group have so many businesses: "they are teachers with a few milk cows on the side, or used clothing salespeople who also keep pigs, or farmers who also raise cassava, matoke, and chickens!" The Kiva fellow met Margaret who is a furniture materials saleswoman and also has invested, with her last loan, in a broiler chicken business. Furthermore, the fellow is amazed to discover on a visit to Margaret's farm that

> She grows enough food to feed her family, and all any others who might come around. She has enough matoke trees (matoke are the bananas that are used to make the staple food here) on her little plot that if the rains are good, she can sometimes sell her surplus matoke in the markets. She has a number of cows that she uses for milk and a few that she fattens for the meat. She also has a few goats, a few pigs, and a few ducks (Chickens for When the Rains Don't Come, 2009).

The explanation Margaret gave to the Kiva fellow is enlightening for the expected surprise that an outsider (readers/viewers and lenders on Kiva.org) might display at such business decisions:

> She explained to me that these home businesses (her farm, cows, pigs, goats, and ducks) together with her furniture business provide enough money for her daily life. She is able to feed her family, buy a few outfits of clothes (this can be a sign of mid-

dle class, as the poor often only have one outfit and the very poor may not even buy one for the smallest of their children), pay school fees, pay rent on one of the farming plots that she rents, and have a little pocket money (Chickens for When the Rains Don't Come, 2009).

On the other hand:

> The chickens are for when the rains don't come, Margaret explained. The rains in Uganda delayed this year. Both of her farming plots produced very little food. She was unable to feed her family and she had none left over for selling. But, she antic- ipated this and took a loan to start up the chicken business. Because of her prof- its from her new chicken business, she was able to support her family and she became a critical customer for those people who choose to put all their efforts into their farms instead of having any other businesses (Chickens for When the Rains Don't Come, 2009).

Like in many other instances found on Kiva.org, the subjectivities contoured through this storyline prototype become actors in a dramatic play whose stag- ing is mediated and also takes place on Kiva's website, so that lenders become acutely aware of their own contribution to the direction and denouement of the play. The following eloquent account by a Kiva fellow acknowledges this situ- ation in an implicit fashion:

> One day Sulton Kurbonov stopped at the MicroInvest office to make a loan pay- ment and I had a chance to meet with this soft-spoken father of seven. He explained that he had taken out a loan to purchase a color television set and my initial reaction was one of disdain. You're struggling to make ends meet yet you're taking out loans to buy a television?....After some reflection I realized what a patronizing attitude I had taken...I thought back to my experience with Sulton. I judged him because his loan was provided in part because of my efforts. Maybe a little tiny piece of me felt like I somehow owned him or at least his decisions. And perhaps I thought that his poverty was a reflection of his capabilities. That may or may not be true, but the essence of human dignity is the ability to make choices— good or bad. Immanuel Kant said, "morality, and humanity as capable of it, is that which alone has dignity." (Poverty Isn't Always Pretty, 2009)

The subject formation taking place within the journals' content relies often on the tendency to contrast the local knowledge to Western epistemes. Such is the instance where the borrowers' culture is represented as a curious blend of prim- itive/modern. Emphasis is placed more on the unusualness of such hybrid sub- jectivities than on the circumstances underpinning their existence. It is relevant in this testimony of one Kiva fellow regarding her visit to a village in Guatemala:

> Someone asked me how it was that I seemed to have (almost) constant access to the internet AND no indoor running water or heat. From an American perspec- tive, it seems irrational and contradictory. But, Guatemala is filled with (seeming) contradictions and contrasts. I suspect that many of my "fellow" fellows have expe-

rienced the same in the countries where they are working. The family I live with
has satellite TV, a wide screen television (and a television in every bedroom) but
they have no indoor running water or heating. They still wash their laundry by hand
in a lavadero, outside. They cook over a wood stove. The water for showers is heat-
ed by a fire lit under a big black drum, which they fill with water early, every morn-
ing (before the water runs out). They make their own masa from the corn that they
grow. And they all have cell phones, MP3 players and their favorite "novellas" (soap
operas) on television (You Have Internet, but No Running Water?, 2009).

Furthermore, in one of the responses to this post, the poster displays with irony
verging on derision and contempt what s/he, most probably a "First" world citizen, easily and uncritically perceives as just "mindboggling," namely, the status
quo of developing societies, where Western modernity/development in the
form of new technologies ("mobile phones and cable television") is to be found
co-existing with the Other's "pre-modernity" in the material form of "crumbling
houses," "dirty dhotis," and "holy robes":

> I live in India and it's much the same. Even after being here for several years, my
> mind boggles at the sight of labourers in filthy dhotis chattering away into mobile
> phones and crumbling houses with cable television. Even the Tibetan monks in
> Dharamsala all seem to have a string around their necks to hold their mobiles in
> place. I guess their holy robes don't have pockets…(You Have Internet, but No
> Running Water?, 2009)

Unfortunately, without an articulate and informed contextualization of such
"mindboggling" material, observations like this one readily acquire critical overtones rebuking "Third World" societies for clinging to backwardness even in the
blatant presence of progress and modernity brought by the West. This fosters
the persistence of ideologies emphasizing the clash of cultures/civilizations, rein-
stating thus the rhetoric of Otherness.

In another instance, a borrower's profile constructs the irrational character
of the "Other," by emphasizing the superstitious attitude of a borrower from
Mozambique:

> He was able to build a brick townhouse and will build another house when he is
> no longer married. He does not want to live with a new wife in the house he built
> with his ex-wife with whom he is separated because she contracted a mental dis-
> ease two years ago (Lourenco Rafael Vilanculos).

A rare instance to be found on Kiva.org website, the use of first person narrative within the profile of borrowers breaks the pattern which describes their *representation* (the third person narrative usually employed in the profiles of
borrowers might easily pass through hegemonic ideological filters at play on
Kiva.org). Only one such example was identified during the website analysis for
the current study, but there might be others like this. Balqees Serwat, a paper

envelopes maker from Lahore, Pakistan, introduces herself for possible Kiva lenders:

> My name is Balqees Serwat. I have six children, all of whom are enrolled in school. My husband and I make paper envelopes. We both have been working very hard throughout our lives to realize our dream of educating all our children. Now we are in a position where we do not know how we will manage to send our children to university, as it is very expensive. Before this time comes, we want to make our business generate a better income and are seeking a loan $400 to purchase more materials to sell in the market (Balqees Serwat).

Lenders' profiles, both verbal and visual contents, reveal assumptions underlying lenders' decisions to make loans, their perspectives on Kiva.org, and, implicitly, their take on the more general development paradigm. Sometimes these assumptions unearth conservative views and beliefs in regards to the binary developed/underdeveloped world and to poverty alleviating practices. Meagan, an Admission officer from Canada, talks about her motivation to be a lender with Kiva.org: "I'm finally at a place where I am able to, and I'm conscious of how I choose to spend my disposable income." (Kiva Lender: Meagan) Katie, another lender, a creativity blogger from the U.S., tells the readers that:

> I love the idea of microloaning and I particularly appreciate how Kiva tells me about each entrepreneur and allows me to select someone to help...Kiva represents a fantastic example of a fresh idea, because it makes microlending approachable for everyone, and connects lenders with the people they're helping (Kiva Lender: Katie).

More directly, as in Katie's example, or less directly, as in Meagan's, an ideology of controlled and selective development praxis takes shape, somehow overshadowed by the networking feature offered by Kiva, where, as Katie argues, lenders are connected to those they're lending to. The Western subject is more easily persuaded to lend when s/he is offered the opportunity and means to oversee the flow of the money even in the context of the greater ideal to alleviate poverty and to contribute to the development of the "Third World." This begs the question whether the network cultural technologies (in this case the Web 2.0 tools, the Internet) do not re-establish the old Western hegemonic practices where the "civilizing" / "saving" enterprise of colonization was pretext for influence and control over the "uncivilized" nations.

Visuals are also revealing of lenders' assumptions and convictions regarding the purpose of their acts of lending on Kiva.org. For instance, the prospect of alleviating poverty is powerfully suggested by the photo posted by Connor, a U.S. student, in his profile page. In the profile photo, he and another young woman, most probably a Westerner like him, are posing at the back of a group of children wearing white shirts, in a play yard. In the background there is a glimpse of a building with a blue entrance door on top of which the viewer can spot some inscriptions, though illegible given the distance at which the photo was shot. All these visual elements might suggest they are pupils in a school that is possibly sponsored by foreign aid. More interesting for the analysis is the fact that the focal position in the economy of the picture layout is occupied by the two Westerners. The choice of the profile photo is not to be discarded as accidental, since it can reasonably represent Connor's statement. He introduces himself on Kiva network as a Westerner lending a helping, charitable, even missionary, hand to the poor nations of the world. The fact that there are no other details, such as the name of the location where the photo was taken, can stress a belief in a universal character of charity and missionary activity proposed for Western youth, possibly a powerful message passed on to other Kiva.org young viewers (Kiva Lender: Connor).

There are nevertheless instances of lenders' voices which acknowledge the historical fact of imperialism and, in a hopeful fashion, liken microlending activities on Kiva to contemporary acts of justice to these historical inequities. Tyler, a substitute teacher from Madison, Wisconsin, argues that he lends on Kiva.org because "this is a great way to send some interest free capital back to people living in areas that our brothers, fathers, and forefathers have been extracting capital…and for other more humanitarian reasons too." (Kiva Lender: Tyler)

CONCLUSION

The analysis of Kiva.org discursive practices cultivated by Internet tools and specifically by means of online network technologies is not meant to discredit its mission and its visible outcomes in the microlending domain; instead, it attempts to complicate and problematize the place and role of ICT within a development paradigm influenced by social networking. Ideally, Kiva.org, through online networking, fosters mobility of representations of subjectivities. At the same time, as Kaplan (2002) asserts, "the value placed on mobility in representations of subjectivity in cyberspace or new technologies is not new, but can be seen to be the full articulation of something old: travel." (p. 33) Travel is of vital significance for the Western culture, as it "produces the self, makes the subject through spectatorship and comparison with otherness." (p. 33) Nevertheless, the travelers enjoying mobility in representing themselves and otherness, i.e., the lenders, are but a segment among all the social actors apparent in Kiva.org networking discourse.

All these move us away from previous conceptualizations of online formations as "community." At the same time, they allow for the formation of ideal individualized global subjectivities that work to the everyday rhythm of seemingly deterritorialized networks. Most importantly, these networks are controlled through (in)visible and implicit hierarchical codes situated in transnational corporate logics.

ACKNOWLEDGEMENT

The authors of this chapter wish to thank Dr. V. Gajjala, Ms. Samara Anarbaeva and Mr. Franklin Yartey for work on related articles that feeds into some of our observations in this paper.

REFERENCES

Alcoff, L. & Potter, E. (1993). Introduction: When feminisms intersect epistemology. In L. Alcoff & E. Potter (Eds.), *Feminist epistemologies* (pp. 1–14). New York: Routledge.

Alcoff, L. (1992). The problem of speaking for others. *Cultural Critique 36*(11), 5–32.

Amelia Ubilluz Mejia's Group. *Kiva.org*. Retrieved April 14, 2009, from http://www.kiva.org/app.php?page=businesses&action=about&id=40584

Balqees Serwat. *Kiva.org*. Retrieved April 18, 2009, from http://www.kiva.org/app.php?page=businesses&action=about&id=19162

Blair, K., Gajjala, R., & Tulley, C. (Eds.). (2008). *Webbing cyberfeminist practice: Communities, pedagogies, and social action*. Hampton Press.

Bob, C. (2005). *The Marketing of rebellion: Insurgents, media and international activism*. Cambridge University Press.

Boyd, D. M., & Ellison, N. B. (2007). Social network sites: Definition, history, and scholarship. *Journal of Computer Mediated Communication 13*(1).

Carroll, T.F. (1992). *Intermediary NGOs: The supporting link in grassroots development*. Kumarian Press, Inc.

Castells, M. *(1996). The rise of the network society, The information age: Economy, society and cul-*

ture, I, Cambridge, MA; Oxford, UK: Blackwell Publishers.

Castells, M. *(1997)*. *The power of identity, The information age: Economy, society and culture, II*, Cambridge, MA; Oxford, UK: Blackwell Publishers.

Castells, M. (2007). Communication, power, and counter-power in the network society. *International Journal of Communication, 1*, 238–266.

Chickens for When the Rains Don't Come (2009). *Kiva stories from the field*. Retrieved April 20, 2009, from http://fellowsblog.kiva.org/2009/04/13/chickens-for-when-the-rains-don% E2%80%99t-come/

Clarke, K. M. (2004). *Mapping Yoruba networks: Power and agency in the making of transnational communities*. Duke University Press.

Enteen, J. (2006). Spatial conceptions of URLs: Tamil eelam networks on the World Wide Web. *New Media & Society, 8(2)*, 229–249.

Gajjala, R. (2004). *Cyber Selves: Feminist ethnographies of South Asian women*. Walnut Creek, CA: AltaMira Press.

Gajjala, R., & Altman M. (2006). Producing cyber-selves through technospatial praxis: Studying through doing. In P. Liamputtong (Ed.), *Health Research in Cyberspace: Methodological, practical and personal issues*. Nova Publishers.

Giddens, A. (1991). *Modernity and self-identity: Self and society in late Modern Age*. Stanford University Press.

Hine, C. (2000). *Virtual ethnography*. Thousand Oaks, CA: Sage Publications.

Kaplan, C. (2002). Transporting the subject: Technologies of mobility and location in an era of globalization. *PMLA*, 32–42.

Kiva Lender: Connor. *Kiva.org*. *Loans that change lives*. Retrieved April 21, 2009, from http://www.kiva.org/lender/connorb

Kiva Lender: Katie. *Kiva.org*. *Loans that change lives*. Retrieved April 28, 2009, from http://www.kiva.org/lender/katiekonrath

Kiva Lender: Meagan. *Kiva.org*. *Loans that change lives*. Retrieved April 26, 2009, from http://www.kiva.org/lender/meagan9783

Kiva Lender: Tyler. *Kiva.org*. *Loans that change lives*. Retrieved April 27, 2009, from http://www.kiva.org/lender/tyler7579

Kiva Lender: Charles. *Kiva.org Loans that change lives*. Retrieved April 18, 2009, from http://www.kiva.org/app.php?page=lender&action=view&name=charles4234&pageID=3

Kiva Lending Team. *Kiva.org*. *Loans that change lives*. Retrieved April 14, 2009, from http://www.kiva.org/community/viewTeam?team_id=94

Kolko, B., Nakamura, L., & Rodman, G. (2000). Race in cyberspace: An introduction. In B. Kolko, L. Nakamura & G. Rodman (Eds.), *Race in cyberspace* (pp. 1–14). New York: Routledge.

Landow, G. (1992). *Hypertext: The convergence of contemporary critical theory and technology*. Baltimore: Johns Hopkins University Press.

Lourenco Rafael Vilanculos. *Kiva.org*. Retrieved April 19, 2009, from http:// www.kiva.org/app.php?page=businesses&action=about&id=66316

Miller, D., & Slater, D. (2000). *The Internet: An ethnographic approach*. Oxford: Berg.

Nakamura, L. (2002). *Cybertypes: Race, ethnicity, and identity on the Internet*. New York: Routledge.

Poverty Isn't Always Pretty (2009). *Kiva stories from the field*. Retrieved April 15, 2009, from http://fellowsblog.kiva.org/2009/04/08/poverty-isnt-always-pretty/

Rheingold, H. (1991). *Virtual Reality*. New York: Touchstone.

Soja, E. W. (1996). *Thirdspace: Journeys to Los Angeles and other real-and-imagined places*. Cambridge, MA; Oxford, UK: Blackwell Publishers.

Terranova, T. (2004). *Network Culture: Politics for the Information Age*. London, UK: Pluto Press.

Tubella, O. (2006). Television and Internet in the construction of identity. In M. Castells & G.

Cardoso (Eds.), *The network society: From knowledge to policy* (pp. 257–269). Washignton, D.C.: Center for Transatlantic Relations.

Wood, A. F., & Fassett, D. L. (2003). Remote control: Identity, power, and technology in the communication classroom. *Communication Education, 52*(3/4), 286–296.

You Have Internet, but No Running Water? (2009). *Kiva stories from the field.* Retrieved April 20, 2009, from http://fellowsblog.kiva.org/2009/04/16/a-study-in-contrasts-2/

CHAPTER SIX

From Howard Dean to Barack Obama

The Evolution of Politics in the Network Society

MICHAEL GIARDINA

The Internet community is wondering what its place in the world of politics is. Along comes this campaign to take back the country for ordinary human beings, and the best way you can do that is through the Net. We listen. We pay attention. If I give a speech and the blog people don't like it, next time I change the speech.

—GOV. HOWARD DEAN (D-VT), DURING THE 2003–2004
DEMOCRATIC PARTY PRIMARY

Were it not for the Internet, Barack Obama would not be president. Were it not for the Internet, Barack Obama would not have been the Democratic nominee.

—ARIANNA HUFFINGTON, WEB 2.0 SUMMIT, 2008

PROEM

February 2003. A full twenty-two months before the 2004 U.S. Presidential election. A time when only hardcore followers of American politics were paying attention to potential Democratic candidates for the highest office in the land (e.g., John Kerry, John Edwards, Dick Gephardt, etc.) give canned speeches on C-SPAN or make the rounds on the cable news channels. Yet, in the eyes of many observers, none of them had yet to say or do anything especially noteworthy to draw a sizable public over to support their respective campaigns (save for having a "D" for "Democrat" after their names). This was a time, you will recall, when George W. Bush, still a few weeks away from launching his unilateral invasion and occupation of Iraq, was riding sky-high in the polls. The events of 9/11/01 hung fresh in the air, driven home ever more problematically by a constant barrage of (questionable) "terror alerts" and quickly receding civil liberties. To make matters worse, the opposition party (i.e., the Democratic party) had, in the 2002 mid-term elections held just a few months prior, seen its Republican counterpart solidify firm majorities in both houses of Congress. These were wilderness years to be sure, a time when Democrats were not only in the minority, but were in many cases voting along with President Bush on issues ranging from massive tax cuts to aggressive foreign policy (e.g., Iraq). For many on the Left, hope did not seem to spring eternal from the choices that stood on the horizon.

And then, in a speech given at the Democratic National Committee Winter Meeting on February 21, a little-known five-term governor from Vermont, Howard Dean, stepped to the podium and turned the wider political establishment on its head when he started his speech by stating bluntly:

> What I want to know is why in the world the Democratic party leadership is supporting the President's unilateral attack on Iraq.

> What I want to know is why are Democratic party leaders supporting tax cuts. The question is not how big the tax cut should be, the question should be can we afford a tax cut at all with the largest deficit in the history of this country.

> What I want to know is why we're fighting in Congress about the Patient's Bill of Rights when the Democratic party ought to be standing up for health care for every single American man, woman, and child in this country.

> What I want to know is why our folks are voting for the president's No Child Left Behind bill that leaves every child behind, every teacher behind, every school board behind, and every property tax payer behind.

> I'm Howard Dean and I'm here to represent the Democratic wing of the Democratic party!

And so started the movement that would grow into the larger "Dean for America" campaign, a campaign that in a few short months would launch Dean from having scant national name recognition[1], a few hundred thousand dollars in his campaign war chest, and a staff of seven huddled together in an office in Burlington, Vermont, to by late-Fall of that year leading handily in several national polls, raising a then-record $52 million in campaign contributions (at least half which was directly raised online[2]), and counting nearly 600,000 financial donors, supporters, volunteers, or un/paid staffers among its ranks. *And all of it seemed to happen overnight.*

* * *

As we well know, Howard Dean did not become President of the United States in 2004. Nor did he win the Democratic nomination. In fact, he recorded only one primary victory (his home state, Vermont, several weeks after he had dropped out of the race).[3] However, his impact on the race—*in fact, on the future of politics and campaigning itself*—has been and continues to be immeasurable, for he and his campaign ushered in a new era (or, at the very least, a new dynamic) to the way in which political campaigns were organized and run, both mechanically and philosophically. As Paul Boutin (2004) of *Slate* presciently observed a few days after Dean officially withdrew from the primary race:

> Years from now, the online Deaniacs[4]—with their MeetUps, their blogs, the mailing lists they put us on without asking—will be the defining aspect of Campaign 2004. For a while, their enthusiasm made pollsters and pundits believe that former campaign manager Joe Trippi, by substituting interactive, free-for-all dialogues for top-down, Karl Rove-style messaging, really could use the Net to take back the White House.

It only took two. In the midst of the 2008 campaign season (which essentially lasted for two years, or since the 2006 midterm elections), the Washington establishment looked on as Barack Obama's campaign for change took the generative seeds of Dean's online political organizing and communication strategies to heights unthinkable to many just a few short years earlier. Whether through active maintenance of an Obama Channel on YouTube or the maturation of the political netroots/blogosphere (including must-read heavyweight sites like DailyKos.com, TalkingPointsMemo.com, or FiveThirtyEight.com) to the creation of the my.BarackObama.com social networking interface or its mind-bogglingly successful online fundraising reach (upwards of $50 million per *month* by the end of the campaign, for a total of half a *billion* dollars raised online in the 21 months of his campaign[5]), Team Obama quite literally re-wrote (or, in some instances, made up on the fly) and effectively codified the new rules for campaign politics in one fell swoop.[6]

To wit: As part of the *Washington Post*'s "Clickocracy" series on the intersection of politics and technology, Jose Antonio Vargas (2008) catalogued some of the more impressive metrics germane to the Obama campaign. I quote at length from Vargas, who was given exclusive access to the online department shortly after the election (the numbers reported are in essence the "official" numbers from the Obama campaign; emphases below are mine):

> Obama's e-mail list contain[ed] upwards of 13 million addresses. Over the course of the campaign, aides sent more than 7,000 different messages, many of them targeted to specific donation levels (people who gave less than $200, for example, or those who gave more than $1,000). In total, more than *1 billion e-mails* landed in

in-boxes. ([In 2004], Sen. John F. Kerry had 3 million e-addresses on his list; former Vermont governor Howard Dean had 600,000.)

A million people signed up for Obama's text-messaging program. On the night Obama accepted the Democratic nomination at Invesco Field in Denver, more than 30,000 phones among the crowd of 75,000 were used to text-in to join the program. On Election Day, *every* voter who'd signed up for alerts in battleground states got at least *three* text messages. Supporters on average received five to 20 text messages per month, depending on where they lived—the program was divided by states, regions, zip codes and colleges—and what kind of messages they had opted to receive.

On my.BarackObama.com, or MyBO, Obama's own socnet [social network], 2 million profiles were created. In addition, 200,000 *offline* events were planned, about 400,000 blog posts were written and more than 35,000 volunteer groups were created—at least 1,000 of them on Feb. 10, 2007, the day Obama announced his candidacy. Some 3 million calls were made in the final four days of the campaign using MyBO's virtual phone-banking platform.

Impressive numbers to be sure. And, clearly, a technological sea change was at hand, even for those who remembered the Dean campaign and looked to it as the model—albeit the beta or 1.0 version—of networked politics in the digital age. However, it was not the fact or mere existence of the Internet (or, perhaps more accurately, even the broader constellation of digital tools at hand), *per se*, that revolutionized the way politics could be structured (after all, by 2008, every major Presidential hopeful had a fully-functioning Web/online fundraising presence, at least to some degree or another), but the very philosophical underpinnings of both the Dean and Obama campaigns that allowed them to more fully realize and implement their grassroots visions: "His campaign wasn't *just* a way to get him elected," notes Richard Wolffe (2009), author of the Theodore White-inspired campaign memoir *Renegade: The Making of a President*. "It was a way to bring people together." (p. 141, emphasis mine)

Thus does this chapter look forward by looking back, starting with the early days of the Dean campaign and following through to the end of the Obama campaign, as I seek to illustrate both the digital evolution of networked political articulations and the promise it holds for a radical, progressive democracy (or, at the least, a more engaged citizenry). In so doing, I will look at several defining aspects of each campaign that reveal the transformational, rather than transactional, nature of their respective political philosophies. For Dean, this will mean focusing on his use of online blogs (including his own) and nascent social networking tools like Meetup; for Obama, examining his overall approach to community organizing through technology in general, and MyBO more specifically.

OF MOUSEPADS, SHOE LEATHER, AND HOPE[7]

> This is what our campaign looks like…concentric circles. If I drop a pebble into the center of the pond, the waves ripple out, self-propelled…We're never going to have a huge field staff, and we're not going to pump millions of dollars into TV and mail, so we have to rely on our supporters to carry the message, to create their own ripples. When I send out an e-mail to 25,000 people, we need everyone to drop pebbles into their own ponds, forwarding it on to their listservs so that really we're reaching 100,000 people. That's how we're building this thing. Get it? (Joe Trippi, Campaign Manager, Dean for America)

Let me begin with the following caveat: Howard Dean was not the first candidate to test the online waters. As Kevin Anderson (2004) reminds us, "Wrestler-turned-governor Jesse Ventura used e-mail to fuel his insurgent candidacy and…John McCain raised $2 million online after his victory over George [W.] Bush in the 2000 New Hampshire primary." (p. 1) In fact, Trippi (2005, n.p.) himself has pointed out that:

> [t]he [John] McCain campaign [in 2000] was really the beginning of the Internet effecting politics. McCain's problem was they had a great campaign on the Net, *but the Net wasn't ready*. It just wasn't mature enough. Not enough people had bought books over Amazon.com, or signed up with an ISP, and you didn't have tools like MeetUp, or a lot of the social networking tools we have now. [emphasis mine].

In both of these instances, the campaigns effectively "kept rigid control over their Web presence" (Hindman, 2005, p. 121); it was a top-down communication strategy, not a two-way discussion. Thus, my focus on and marking of the Dean campaign as the starting point of networked politics in the digital age derives from an understanding that the Net was sufficiently mature for the purpose at hand (in effect, an adolescent version of Web 2.0), and that the addition of two-way communication (real or imagined) between citizen and campaign, as well as the forum for citizens to interact with each other in a virtual space, was a defining feature.[8]

Here I should note that by "networked politics" I am deploying the term both in the broad sense of "rethinking political organization in an age of movements and networks," as well as in the more narrow sense of specifically referencing "the new techno-political tools made possible by the revolution in information technology and their potentialities for transformative thought, action, and communication." (Wainwright, et al., 2007, p. 4) Correlatively, the above deployment also assumes that there exists a degree to which 'being connected' to the political arena and/or discourse is now no longer the sole preserve of big-money donors or traditional party activists (as had been the case in past election cycles) but rather 'connected' in the sense of being linked together—in common purpose—through an ever-evolving digital network of blogs, Facebook friends, Twitter tweets, YouTube clips and channels, Google Reader feeds, candidate Web pages, and the like: in other words, a network that "ordinary" citizens could easily "join." Such an open space, or emerging "global collectivist society," if you prefer Kevin Kelly's (2009) more audacious "new socialism" terminology, is here exemplified by "a spectrum of attitudes, techniques, and tools that promote collaboration, sharing, aggregation, coordination, and a host of other newly enabled types of social cooperation (n.p.). As I will detail below, the maturation of such a networked political space is at the heart of the arc from Dean to Obama.

* * *

Much has been made about Dean's two major "What I want to know..." speeches, and how they served to generate increased public awareness of his campaign at a time when a strong oppositional voice in American politics was noticeably lacking, especially to the ears of liberals and progressives. This much is true. However, when we begin to unpack the complex history of the campaign, we find that Dean's online presence—in the form of unofficial blogs, online communities, self-organized offline volunteer efforts, etc.—actually predated his own campaign's official blog by roughly one year. Jerome Armstrong, the founder of MyDD.com who would eventually become part of the official campaign, was one of the earliest supporters to bring an online face to Dean when he began blogging about him on MyDD.com in June 2002. In August 2002, Armstrong, along with fellow Dean supporters Aziz Poonawalla and Anna Brosovic (aka Annatopia), would become bloggers on Poonawalla's DeanNation blog (DeanNation.blogspot.com). Over the next six months, the DeanNation blog became the de facto site on the Web for those looking to follow the Dean campaign (so much so that they began receiving e-mail from individuals who thought they were contacting the *real* campaign headquarters). Other bloggers, such as Matt Yglesias, Matt Singer, and Ezra Klein would also see time as DeanNation bloggers, becoming part of what Armstrong (2007) later referred to as a "farm team of sorts for future liberal bloggers." (p. 44)[9]

The MyDD/DeanNation example is only one of the myriad examples of the early phases of the campaign, wherein a passionate "tribe" (see Godin, 2008) of Dean supporters within the netroots effectively began organizing itself around the candidate and began vocalizing support for him without input from the campaign hierarchy. Zephyr Teachout (2007), who was the campaign's Director of Internet Outreach, recalls one such example in which she approached an existing New York for Dean Yahoo! listserv in the hopes of coordinating an upcoming event; she learned that, on their own and without guidance from the campaign, "they had already built a small existing structure…had formed a New York for Dean committee, and were flyering for Dean at antiwar rallies." (p. 59) When she attempted to task the group with turning out several hundred people to a New York City event, which was to be held the following month, she learned that "They were already planning on it." (p. 59) And how did things turn out for the event in question, which the campaign had practically no hand in organizing, but in which the candidate was going to show up? Teachout (2007) again:

> [T]hey flyered extensively, made hundreds of phone calls, and used emails to get their friends there. They put out the word to enroll the press, and the *New York Times*, the *Village Voice*, and two local TV stations showed up at the event—despite having initially rebuffed our official press secretaries' efforts to get them there.…By the time the governor arrived over 550 people were waiting (p. 60).

As the campaign progressed into the summer months of 2003, events such as the one above became the rule, rather than the exception to it, with the most notable being an event at the Plaza Saltillo in Austin, Texas, in which 3,200 people showed up to hear Dean speak. The campaign's involvement? It sent one mass e-mail to four hundred eighty-one Austin-area supporters who, *on their own*, "became little campaign managers, putting up signs and posting flyers, arranging media, and passing on the word to their friends…[They also] decided, on their own, to leaflet every Latino neighborhood in Austin (Trippi, 2004, p. 116).[10] In effect, these were the sprouting seeds of the open-source or decentralized campaign structure Trippi (2004) had long been visualizing, but only now had the technology to implement. A campaign structure that allowed—in fact *encouraged*—various communities and constituencies to "take basic ideas and adapt them for local conditions, creating solutions that [were] tailored to fit their actual problems and the real collective capacities of their situation." (Berlinguer, 2007, p. 58) Or, in more simple terms, "a willingness to let a decentralized network of supporters play a tactical role" (Wolf, 2004, p. 1) in the shaping of a campaign from the ground up. In this sense, argues Steven Johnson, author of *Emergence: The Connected Lives of Ants, Brains, Cities, and Software* (an early text on digital social networking) the Dean campaign was a techno-political cousin to the open-source Linux operating system, with Dean playing the role of "a system running for President" rather than just an individual (quoted in Rich, 2003, p. 1).

By the time late May 2003 rolled around, the traditional media was starting to take notice of just how differently Dean for America was being run in comparison to the normative, top-down model of political organizing that was utilized by the rest of the declared candidates for President. Ryan Lizza (2003), writing for *The New Republic*, noted the following:

> Some campaign managers devote their energies to working the elite press or courting union leaders or wooing donors. But Trippi seems to spend an inordinate amount of his time checking MeetUp numbers, posting to liberal blogs, sending text messages to supporters who have signed up for the Dean wireless network, and otherwise devising ways to use the Internet to build what Trippi envisions as "the largest grassroots organization in the history of this party." And his efforts might actually be paying off: While many predicted that Dean would fade away once the war was no longer a salient issue,[11] there is little evidence that the former Vermont governor's supporters—originally drawn to Dean when he was forcefully speaking out against war in Iraq—are deserting him. In fact, the Internet might account for Dean's staying power.

At the same time, however, there was a far greater degree of derision than praise directed to the campaign for its Internet strategy and tactics. As Trippi (2008) later remarked: "The majority of the Democratic establishment thought the Internet and the netroots were like a bar scene out of *Star Wars* or a small group of people talking to each other in their underwear. They thought it was a big waste of time, a fad." (p. 9) To give one example, Jim Jordan, then-campaign manager for Sen. John Kerry's team, dismissed Dean's Internet efforts in *Time* magazine (an issue which featured Dean on the *cover*) by stating: "It's like watching my 13-year-old daughter instant-messaging. It's not particularly about politics and policy. It's almost like a reality show." (quoted in Tumulty, 2003)[12]

In the remainder of this section, I want to look at two particular facets of the campaign that directly changed the face of (networked) politics: (1) the official campaign blog for online communication and (2) the use of social networking tools such as MeetUp, among other developments, for offline organizing. These two breakthroughs and their attendant derivations (such as fundraising)—along with the candidate himself, who was by this point engaging in televised debates with the other candidates and appearing more regularly on various mainstream news programs such as NBC's *Meet the Press*—would in part lead *New York Times* Op-Ed columnist Frank Rich (2003) to conclude by the end of 2003 that Dean should *not* be compared to Barry Goldwater or George McGovern (fringe candidates who became serious challengers before losing mightily) but rather that "it may make more sense to recall Franklin Roosevelt and John Kennedy," in the sense that FDR's use of the radio for fireside chats became a political force, and Kennedy's use of the television signaled yet another paradigmatic shift in the nature of politics. In other words, a continuum of Roosevelt:Kennedy:Dean::Radio:Television:Internet.

BLOGGING FOR AMERICA

In retrospect, Blog for America (or BfA, as became the shorthand notation), the proverbial public face of the Dean campaign's Web presence, looks (and, in fact, was) very simplistic when compared to the sleek ocean liner the Obama online team would roll out a few years later. Three columns, a few daily blog entries from campaign staff complete with a comments section under each, a blogroll and the legendary fundraising "bat" icon were all mainstays by the time the campaign moved into the long summer months of 2003. In its earliest iterations, the site contained no glossy campaign videos, Java or Flash architecture, or virtual phone-banking applications that would later dominate the Obama site by the end of 2008. But to those who frequented the site it was a glorious development, "not so much a Web site as a complex, palpable, but elusive entity…[that]…gave the campaign a virtual presence easily mistaken for something tangible, as if a campaign staffed by tens of thousands of ordinary people existed around the clock through a series of organic conversations available at the click of a mouse." (Kerbel & Bloom, 2005, pp. 3–4)

Over the course of the campaign, BfA was home to (or in some cases, was the generative impulse for) a series of events that directly furthered the public perception of Dean as a straight-talking Washington outsider and solidified his candidacy as serious rather than fringe. Notable among these events were:

- A July 28, 2003 virtual fundraising event, suggested by a blog reader, that featured Dean, sitting at his desk in Burlington eating a $3.00 turkey sandwich while the whole thing was streamed live on the Net, and that raised more than $500,000 in two days time (the premise being that Vice President Cheney was aiming to raise $300,000 at a luncheon with one-hundred fifty $2,000-a-plate donors on the same date);
- the implementation of the so-called "bat" icon (similar to a traditional fundraising thermometer), which announced a particular monetary goal, and was updated accordingly (campaigns traditionally never publically announce their fundraising goals, for fear that if they are not met, a campaign will suffer negative publicity);
- and, just as importantly, the overall growing comfort of campaign supporters to blog on the site, as well as elsewhere (especially on DailyKos.com), about Dean, his campaign, and politics in general.

Markos Moulitsas Zuniga, founder of DailyKos.com, which is by most accounts the gold standard when it comes to politically-oriented blog communities[13], remarks that the true genius of the emerging [liberal] blogosphere was that it allowed unfettered conversations to take place at the exact moment when "the media was under-serving people. Someone needed to bypass those media gate-keepers who were dictating what the public consumed and squashing dissent." (p. 8) In the context of the Dean campaign, BfA became in many ways a virtual Habermasian public sphere, a literal realization of a performative place where participants, through their online encounters, "created something new—the possibility of true political life." (Goldfarb, 2008, p. 131) In many ways, contend Matthew R. Kerbel and Joel David Bloom (2005), Blog for America "worked to energize readers for participation in the political process by assuring them that their work [was] meaningful and valuable and that they [were] not alone in their efforts." (p. 4)

Not only did BfA serve to "rally the troops" (and raise money) as it were, but, more importantly, it solidified the idea that the campaign really *was* listening to and engaging in conversation with its supporters on everything from policy questions to community organizing. Trippi (2004) puts it in very direct terms:

> [T]he biggest myth of the 2004 election was that Joe Trippi was managing Howard Dean's presidential campaign. *They* were managing the campaign. It wasn't head-quartered in Burlington; it was *out there*. Anything we could do, *they* could do better (p. 116, emphases in original).

In terms of what "they" were "doing better," it ranged from the smallest details—such as modifying campaign posters and flyers or creating software applications that were specially written for the campaign to use, such as Deanspace and DeanLink (which became the campaign's own in-house version of MeetUp)—to the grandest political notion: that the campaign was actually a *movement* that could "take the country back." In Trippi's view, "we needed to take this thing all the way to open-source, put the code out there and see if the people could improve it." (p. 116) David Weinberger (2003), author of *Small Pieces Loosely Joined: A Unified Theory of the Web*, a key text on Internet connectivity, puts it this way:

> The goal [was] not necessarily to have messages flowing up and down. Democracy is supposed to be about people talking with each other about what matters to them…Ideas from the grassroots don't have to go back up to headquarters to be adopted…The Dean campaign instead [gave] you the tools to institute your own ideas without involving headquarters (quoted in Wolf).

To give but one more example, Nicco Mele (2007), the campaign's webmaster, recalls a story from 2004 that highlights the open-source nature of which Trippi and Weinberger speak:

> Walking through Brooklyn to meet my brother, I noticed a flyer posted on a telephone pole advertising a Dean tabling event at the Brooklyn Farmer's Market. It

invited Dean supporters to come help get the word out. The flyer was well designed and caught the eye, and I marveled at the amazing work and energy people were pouring into the campaign. The next day I flew to Seattle, and walking around downtown Seattle I saw the exact same flyer—but this time advertising a Dean self-organizing event in Seattle, not Brooklyn.

Now I was curious. I went online and discovered that a group on the West Coast—I think it was North Bay for Dean—had developed their own poster tool. It took the details of your event and inserted them into a pretty little flyer, complete with tear-offs for more info. Thousands of people around the country were using this tool to make flyers for their local Dean events, and here I was, ostensibly in charge of the web operations for the entire campaign, and I had only just learned of its existence. It was a beautiful, amazing moment. Some folks in California had seen the need for a clever tool for poster making—and instead of waiting on the campaign to provide something, they went ahead and built it themselves on their own, and then invited the nation to use it (pp. 182–183).

A nice story to be sure. But what is more, stories like the one Mele recounts above were happening *all over the country*, in no small part due to one of the key *offline* organizational aspects of the campaign, but ones that were realized only through the use of *online* resources. Namely, Meetup.com.

MEETING-UP FOR CHANGE

Quick—answer this question: What was online social networking like in 2003? Facebook? Nope, it launched in February 2004, and at that point was limited to Harvard and a few other college campuses. YouTube? Nope, that wasn't launched until February 2005. Twitter? Nope, it wasn't even founded until March 2006. Even Friendster, the first social networking site, only went live in March 2003. All of these major sites played a significant role in the Obama campaign, yet none were available to Trippi & Co. as they stormed through the summer of 2003. What they did have available, however, and what eventually became a key cog in the Dean Machine, was a social networking portal called MeetUp.com, which allows *online* users with similar interests (e.g., the Los Angeles Lakers, wine tasting, vintage cars, etc.) to meet-up *offline* at a location designated by the site (usually a coffee shop).

In typical fashion, the campaign became aware of Meetup not of its own volition, but from a voice outside the campaign (in this case, Jerome Armstrong, who recalls that a Meetup official named William Finkel reached out to members of the aforementioned DeanNation blog in January 2003 to see if there would be any interest utilizing the service to organize for Dean [2007, p. 47]). While Armstrong lobbied Trippi about the relative merits of Meetup, Aziz Poonawalla wrote on the DeanNation blog:

> Meetup.com has a Meetup page devoted to Howard Dean, which will be enormously useful to Dean supporters who want to organize….It's really up to the campaign whether it wants to give an official blessing to these kinds of netroot efforts or not—but they will likely happen regardless. Still, with some explicit outreach, the campaign can reap far greater rewards (quoted in Armstrong, 2007, p. 47; see also the original blog entry at http://dean2004.blogspot.com/2003/01/netroots-are-still-more-organized-than.html).

Within two weeks of this post, there were 432 people signed up for Dean Meetups; a small number, to be sure, but the most of any of the other candidates; an impressive feature given Dean's relative obscurity (Trippi, 2007). After the campaign put a Meetup link on its homepage, the Meetup for Dean numbers jumped quickly, up to 2,700 in a week or two, then 8,000, and continuing at a rapid rate until finally ending with an astonishing 190,000 by the end of the campaign season.

Importantly, the campaign did not just see these 190,000 individuals as numbers on a spreadsheet; that is, supporters to whom they could solicit campaign contributions. Rather, these ordinary citizens became active members of the Dean *movement*. In Trippi's (2007) words, Meetup was another contribution to "the process of asking the American people to participate once again in their common future…to come together in common cause—in town meetings and town halls—to forge a new American Century from the bottom up." (p. 125)

Even though the campaign retained some level of involvement (including data analyses of what was happening at the events), and in fact provided material to Meetup leaders later in the process (such as stickers, campaign leaflets, or a DVD to be played at the event), the onus was on each Meetup group to actually follow through. In many ways, this was traditional grassroots organizing at its best—*people talking with their friends and neighbors face-to-face about the issues of the day, and then doing something tangible to bring about change*—with the only real difference being the communicative tools that brought them together in the first place.

* * *

In the last analysis, the legacy of the Dean campaign really is not, as some might have it, about its use of technology or technological advances *per se*, but rather how such advances were incorporated into and facilitated a shift in the very way political campaigns could be run. Dean (2004) himself is clear on this point, arguing that although small donor programs, house parties, interactive websites, and the like were copied by every other candidate, the most critical piece—and the one that no one dared copy—was "ced[ing] power to local folks and let[ting] them run things in their areas as they saw fit." (p. 155)

By the time the 2008 campaign season would come around, Web 2.0 would be in full swing, but in order to truly capitalize on it, someone would have to embrace that most "critical piece"—"a cognitive change in those who run campaigns." Ergo, Barack Obama.

EARLY REFLECTIONS ON OBAMA'S ONLINE MACHINE

> It was a masterclass in political campaigning, a high-water mark. They built on the lessons from Howard Dean, and let people build their own networks. my.BarackObama.com was inspired.
>
> —MARK FLANAGAN, HEAD OF STRATEGIC COMMUNICATIONS
> FOR THE BRITISH GOVERNMENT (QUOTED IN KISS, 2008)

If the biggest myth of the Dean campaign of 2004 was that technology and the Internet alone were the main reasons he was able to raise so much money, than the biggest myth of the Obama campaign was that technology and the Internet alone were the main reasons he was able to effectively organize and manage so many volunteers throughout the decentralized campaign structure. However, as Blue State Digital's Thomas Gensemer (2009) explains,

> The biggest thing is that [Obama] was a community organizer 20 years ago, and he knew that you as my neighbor knocking on my door meant more than a paid organizer or even himself knocking on the door. Obama saw technology as the only way to transfer traditional community organizing to a national level, with volunteers and donors signing up online and then being encouraged to go out to recruit further volunteers, hold meetings and house parties, spread the message (quoted in Talbot, 2009).[14]

Ah yes, community organizing. Or, the same thing Gov. Sarah Palin (R-AK) mocked in her speech accepting the Vice Presidential nomination at the Republican National Convention. But as Trippi would say, she just didn't get it. Like with Dean, the Obama campaign was not about technology, per se: It was about organizing a grassroots network of supporters on a scale never before achieved: "technology wasn't *just* a tool in the arsenal, but a transformative force…[that formed]…the biggest nationwide field organization ever created." (Franklin-Hodge, quoted in Talbot, 2009, emphasis mine)

Central to this effort was, of course, Obama's inherent understanding of the paradigm at hand. For example, Facebook's[15] Marc Andreessen notes that, "Other politicians I have met with are always impressed by the Web and surprised by what it could do, but their interest sort of ended in how much money you could raise. [Obama] was the first politician I dealt with who understood that the technology was a *given* and that it could be used in new ways." (quoted in Carr, 2008, emphasis mine) Likewise, Daniel J. Weitzner (2008), co-director of the Decentralized Information Group at MIT's Computer Science and Artificial Intelligence Laboratory and a member of the Obama campaign's Technology, Media, and Telecommunications committee, concurring with Andreessen, explains, "There's an appreciation for the Internet, a recognition of the transformative value of the Internet that I think will go a long way toward shaping the approaches that Sen. Obama would take as president." (quoted in "Is Obama a Mac and McCain a PC?", 2008)

With the above as background, and informed by the previous discussion of the Dean effort, the remainder of this section will look primarily at MyBO, the social networking site hosted by the Obama campaign, and its impact in organizing a legion of supporters to not only spread the word about his candidacy, but also become active participants who were invested in the campaign.

HI-HO, HI-HO, IT'S OFF TO MYBO WE GO!

Everything I see reflects the community organizing experience. I see the consensus-building, his connection to people and listening to their needs and trying to find common ground. I think at his heart Barack is a community organizer. I think what he's doing now is that. It's just a larger community to be organized.

—REV. ALVIN LOVE, 2007

Change will come from a mobilized grassroots.

—BARACK OBAMA, 1995, DREAMS FROM MY FATHER

Wired's Sarah Lai Stirland (2008), commenting on the Obama campaign's organizational efforts the week prior to the election, summed up the end-point result in this fashion:

Obama is the first to successfully integrate technology with a revamped model of political organization that stresses volunteer participation and feedback on a massive scale, erecting a vast, intricate machine set to fuel an unprecedented get-out-the-vote drive in the final days before Tuesday's election. [The campaign had] 19,000 "neighborhood teams" [in Florida] as of late October, focused on 1,400 neighborhoods across the state, according to a recent report from the St. Petersburg Times. The teams are directed by about 500 paid campaign field organizers, and are replicated nationally. In all, the Obama campaign estimates that 1.5 million volunteers are helping it to get out the vote in the battleground states.

The technological core of these organizational efforts can be traced to my.BarackObama.com—or MyBO, for short—the in-house social networking site that enabled end-user virtual phone-banking, pro-active fundraising, scheduling of local events, data base entries, blogging, neighbor-to-neighbor canvassing, and a host of other features. Not surprisingly, David Talbot (2008) of

Technology Review reveals that the MyBO tools, were, at their core, "rebuilt and consolidated versions of those created for the Dean campaign." The key difference, at least architecturally, was that while the Dean campaign either developed or fractionally implemented most of the tools the Obama camp eventually used, such as SMS, phone tools, Web capacity, online fundraising, etc., the Obama campaign, via Blue State Digital, "did a lot of nice work in taking this crude set of unrelated applications and making a complete suite," akin to Microsoft's *Office*, Apple's *iLife* integration, or even Google's Wave (Teachout, in Talbot, 2008).

"Rather than merely joining this network, passively clicking a button to donate or express an allegiance to Obama," offered Gensemer, "[M]embers were encouraged to go out into the real world to knock on doors, hand out leaflets and spread the word. The site then encouraged these efforts to be recorded and shared with the online community, making the user feel empowered and on the front line of the campaign." (quoted in Moss & Phillips, 2008) Thus, the goal of the interface was the activation of volunteer and supporter energy under the auspices of campaign tasks: "[T]he point of the campaign is to get someone to donate money, make calls [*ed*: some 3 million such calls in the last four days of the campaign alone], write letters, organize a house party. The core of the software is having those links to taking action—to doing something." (Franklin-Hodge, in Talbot 2008) Even though much of the Obama campaign was in fact structured hierarchically rather than horizontally (and perhaps not as free-wheeling and open-sourced as was the Dean campaign), the actual execution of it depended just the same on the people who were "out there" engaged in a movement for change.[16] As Joe Erwin (2008) wrote in *AdvertisingAge*, this was *not* traditional advertising; it was "building a brand the way social networks are built out. And perfect strangers [were] talking to each other over the Internet and sharing stories about why, for the first time in their lives, they [were] signing on to a campaign website and donating money."

A correlative element in the MyBO experience separate from strictly organizing principles was the incorporation and circulation of some 1,800 official YouTube clips of Obama speeches, interviews, and campaign events, all of which could be quickly and easily shared via e-mail and personal Web pages. The most significant video clips were undoubtedly his much-heralded speech on race ("A More Perfect Union") that he delivered in the aftermath of controversial comments made by his former Pastor, Rev. Jeremiah Wright (which, ironically, first surfaced publicly on YouTube), and the now-famous will.i.am music video title "Yes We Can." In the first instance, the 37-minute speech, which likely saved the campaign at a time in the primaries when Obama was vulnerable (see especially Wolffe, 2009), was re-viewed by tens of millions online, helping as much to spread the word about Obama as quelling fears among those who were troubled by his association to Wright. In the second instance, the video by Black Eyed Peas' frontman will.i.am, which featured numerous celebrities voiced over by excerpts from an Obama speech, became an instant Internet sensation that generated upwards of one-million views per day in the first several weeks of its release, as well as buzz in the traditional media.

Taken together, the sum total of the Obama campaign's direct engagement with volunteers and potential voters through new media communication channels ultimately resulted in transforming the campaign from being "just another politician" (albeit a gifted one) into one whose brand image approached that of what Douglas Holt (2004) would call a "cultural icon," and enabled him to more effectively operate as a pragmatic "cultural activist" who appeared sensitive to the contextual and material realities of the 2008 election season (i.e., end of Bush era, war in Iraq, unsettled economy, prominence of technology in society, engaged citizens, etc.), and who then leveraged that cultural knowledge into the groundswell of support he ultimately received. In this way, the Obama brand was, in the parlance of digital-advertising agency Resource Interactive, an "OPEN Brand," an acronym for an *O*n-Demand, *P*ersonal, *E*ngaging, and *N*etworked Brand wherein "not only [did] people feel they [knew] who he was, they [felt] trusted to share their views. And they [got] constant feedback from the campaign and each other." (Scholl, in McGirt, 2008) What this accomplishes, both with respect to Obama and more generally as a philosophy, is reinforcing "the notion that everyone is included and that this movement is actually a conversation to which everyone is invited." (McGirt, 2008)

CODA

> So in four years, the technology and know-how skipped Mercury and Gemini and went straight from the Wright brothers to the Apollo program....I think we're about to see the first wired, connected, networked presidency.
>
> —JOE TRIPPI, 2008, DISCUSSING THE TECHNOLOGICAL EVOLUTION OF THE DEAN TO OBAMA CAMPAIGNS

Two campaigns. Two very different results. Yet both remain inextricably linked to each other through the use of new media and social networking to advance a discussion about enacting transformational change in the world, from the ground-up, by ordinary people. Because Dean did not succeed in ascending to the highest office in the land, his strategy was mocked or at the least looked toward with trepidation. In other words, even though it wasn't a case of "Well, the strategy failed so therefore the strategy was *wrong*," when it came time for 2008, Andrew Rasiej, co-founder of *techPresident*, recalls, "As far as major political circles were concerned, Howard Dean failed, and therefore the Internet didn't work." This is the sort of narrow-minded thinking that resulted in every serious candidate save for Obama (and later, to an extent, Hillary Clinton) treating his or her campaign Web site as an ATM machine, a fundraising portal, and nothing much more. Obama, building on and nearly perfecting the campaign tools and philosophy first crafted in its Web 1.0 iteration by Dean, had a leg-up in organizing from the start (even if it took a while for all of the pieces to come together and translate into tangible results).

And now that he is in office, President Obama has, in theory, a legion of passionate supporters, still connected through various online social networks, ready to mobilize in their communities to support his agenda (for example, as was seen with respect to the passage of the massive stimulus package in the first few weeks of his administration). He is also continuing to use sites such as YouTube to broadcast his weekly radio address (which we might as well call his weekly Vlog), and has moved toward creating more transparency in government by posting online drafts of bills and streaming his Health Care summit on the White House's redesigned homepage; and offering a place for citizens to ask questions of administration officials (who have taken to responding to such questions with Vlogs of their own).

Let it be said that while Howard Dean may not have won his Presidential primary, his (and Trippi's, and his early supporters') idea for an open-source political future most assuredly *has* won (at least, for this round). In the last analysis, it is this singular contribution that may be of far greater consequence for the future of politics—in fact, *for the very future of this nation*—than any legislative victory he may have hoped to win as President. And for that, those of us committed to the ideas of an open and free democratic society owe him a debt of gratitude.[17]

NOTES

1. In several early nationwide polls that Spring, Dean would fail to register even a percentage point of support, oftentimes having an asterisk next to his name indicating "did not register support."
2. Hindman (2005), crunching the numbers, speculates that Dean would have raised approximately $11 million (or just 21 percent of his $52 million take) without the Internet.
3. Dean won exactly as many states as did every other candidate in the race not named John Kerry: one. He dropped out after finishing third in the Wisconsin primary on February 17.
4. "Deaniac" was a term coined to refer to a supporter of Howard Dean, especially as related to someone who interacted with the campaign in the online environment.
5. The final numbers, according to the *Washington Post*, broke down as follows: "3 million donors making a total of 6.5 million donations online adding up to more than $500 million. Of those 6.5 million donations, 6 million were in increments of $100 or less. The average online donation was $80, and the average Obama donor gave more than once."

(Vargas, 20 November, 2008)

6. Of note, Joe Rospars, a 27-year-old veteran of the Dean campaign's Internet department—especially its landmark Blog for America—was the Director of Obama's new media department and reported directly to campaign manager David Plouffe (Vargas, 2008).

7. "Of mousepads, shoe leather, and hope" was a familiar refrain during the Dean campaign to describe the core of the online movement. For what I consider to be the two most useful statements on the Internet's role in Dean for America, see the book of the same name by Zephyr Teachout and Paul Streeter (2007), as well as Joe Trippi's (2004) pseudo-campaign memoir *The Revolution Will Not Be Televised: Democracy, the Internet, and the Overthrow of Everything*. Both of these texts were invaluable sources of information for this section.

8. Additionally, it is worth noting that the Internet had been effecting American political discourse since the late-1990s, most especially Matt Drudge's DrudgeReport.com, which played a key role in publicizing, albeit with a conservative bent, the Bill Clinton-Monica Lewinsky developments. For purposes of this chapter, I have limited the discussion to the organization of political campaigns and their off-shoots, e.g., discussions on candidate-specific blogs such as the ones discussed below.

9. For example, Klein is now an Associate Editor at *The American Prospect*, and Yglesias was an Associate Editor with *The Atlantic* through 2008 before moving on to become a Fellow at the Center for American Progress Action Fund.

10. Teachout's (2007) chapter is an invaluable resource concerning the Dean Internet machine.

11. Saddam Hussein had been captured in Iraq shortly before Lizza wrote those words about the saliency of the "war" argument/platform.

12. How quickly times change. Roughly six years later, President Obama was releasing his weekly radio address as YouTube video clips, and Republican Senators were responding with 160-character tweets on Twitter.

13. DailyKos currently averages 800,000 site visits per day. It reached its apex of visitors in October 2008 (the month before the 2008 Presidential election), when it recorded 75.5 million site visits (or roughly 2.5 million per day). In the spirit of full disclosure, I was a somewhat regular participant on this blog during 2003–2005, with a chronologically-assigned User ID number under 500 (there are now 100,000-plus registered members).

14. Which brings us back to Dean once more: "The too-often untold story of the [Obama] campaign is that the work that we were able to do in 2007 and 2008 actually started in 2005 and 2006 with Dean at the Democratic National Committee" (Gensemer in Moss & Phillips, 2009).

15. Facebook co-founder Chris Hughes served as a member of the campaign leadership team (and contributed to the build of MyBO), as did Google CEO Eric Schmidt and other Silicon Valley executives.

16. Other technological innovations or implementations included targeted SMS (text messaging) alerts, including three on Election Day, and a free iPhone application that organized contact lists into state-by-state categories and included campaign position papers and statements to aid in door-to-door canvassing.

17. Despite their technological marriage, Dean and Obama have not seen eye-to-eye throughout the first year of the Obama presidency, especially on issues related to health care. Dean,

as standard-bearer of the progressive netroots, has publicly butted heads with Obama and been looked upon with such scorn by other figures in the Washington political establishment (such as White House Chief of Staff Rahm Emanuel) that Steve Grossman, former National Chairman of the Democratic National Committee (1997-1999) has stated: "You know the expression, to be a prophet without honor in your own land? That's Howard Dean."

REFERENCES

Ambinder, M. (15 June, 2009). The revolution will be Twittered. *The Atlantic Monthly*. Retrieved 15 June 2009 http://politics.theatlantic.com/2009/06/its_too_easy_to_call.php

Anderson, K. (2004, 14 January). Internet insurgent Howard Dean. *BBC News*. http://news.bbc.co.uk/2/hi/americas/3394897.stm

Armstrong, J. (2007). How a blogger and the Dean campaign discovered each other. In Z. Teachout and P. Streeter (Eds.), *Mousepads, shoe leather, and hope: Lessons from the Howard Dean campaign for the future of Internet politics* (pp. 38–54). Boulder, CO: Paradigm.

Berlinguer, M. (2007). Open source for the operating systems of the earth: A metaphor for new institutions? In H Wainwright, O. Reyes, M. Berlinguer, F. Dove, M. Fuster i Morrell, & J. Subirats (Eds.) *Networked politics: Rethinking political organisation in the age of movements and networks* (pp. 57–65). Transnational Institute, Amsterdam.

Boutin, P. (18 February, 2004). Howard's web: The Internet couldn't save Dean, but it could still help Kerry. *Slate*. Retrieved 13 June 2009 from http://www.slate.com/id/2095707/

Carr, D. (10 November 2008). How Obama tapped into social networks' power. *New York Times*. B1.

Erwin, J. (27 February 2008). How they grew Brand Obama. *AdvertisingAge*. http://adage.com/campaigntrail/post?article_id=125377

Giardina, M. D. (in press a). Toward a politics of hope: Performing political reality in the age of Obama. *Cultural Studies/Critical Methodologies*.

Giardina, M. D. (in press b). Barack Obama, religious identity, and the mediated spectacle of the 2008 U. S. Presidential election. In C. Stonebanks, S. Steinberg, and J. L. Kincheloe (Eds.), *Teaching against Islamophobia*. New York: Peter Lang.

Godin, S. (2008). *Tribes: We need you to lead us*. New York: Portfolio.

Goldfarb, J. C. (2008). *The politics of small things: The power of the powerless in dark times*. Chicago: University of Chicago Press.

Hindman, M. (2005). The real lessons of Howard Dean: Reflections on the first digital campaign. *Perspectives on Politics, 3*(1), 121–128.

"Is Obama a Mac and McCain a PC?" (2008, 21 March. *Los Angeles Times*. *http://latimesblogs.latimes.com/technology/2008/05/is-obama-a*-mac.html

Kelly, K. (22 May, 2009). The new socialism: Global collectivist society is coming online. Wired. Retrieved 16 June 2009 from *http://www.wired.com/culture/culturereviews/magazine/17-06/nep_newsocialism*

Kerbel, M. R., & Bloom, J. D. (2005). Blog for America and civic involvement. *The Harvard International Journal of Press/Politics, 10*(3), 3–27.

Kiss, J. (2008, 10 November). Why everyone's a winner. *The Guardian* (London), p. 3.

Kriess, D. (2008). Taking our country back: The new left, Deaniacs, and the production of contemporary American politics. Paper presented at the *Politics: Web 2.0* conference, Royal Holloway, University of London, April 17–18.

Lizza, R. (21 May, 2003). Dean.com. *The New Republic Online.* Retrieved 20 November 2008 from http://www.tnr.com/doc.mhtml?pt=MsFThRpEfeVRNVf9ei9j2

McGirt, E. (2008, 19 March). The brand called Obama. *Fast Company.* *http:// www.fastcompany.com/magazine/124/the-brand-called-*obama.html

Mele, N. (2007). A web activist finds Dean. In Z. Teachout and P. Streeter (Eds.), *Mousepads, shoe leather, and hope: Lessons from the Howard Dean campaign for the future of Internet politics* (pp. 179–191). Boulder, CO: Paradigm.

Moss, S., & Phillips, S. (18 February 2009). Is this man the future of politics? *The Guardian* (London), p. 12.

Rich, F. (21 December 2003). Napster runs for President in '04. *New York Times.* Available http://www.nytimes.com/2003/12/21/arts/21RICH.html

Russell, A., & Echchaibi, N. (Eds). (2009). *International blogging: Identity, politics, and networked publics.* New York: Peter Lang.

Salant, J. D. (2000, February 2). Fund-raising champion Bush lags online. *Boston.com* Available *http://graphics.boston.com/news/politics/campaign2000/news/Fund_raising_champion_-* Bush_lags_online.shtml

Silberman, M. (2007). The Meetup story. In Z. Teachout and P. Streeter (Eds.), *Mousepads, shoe leather, and hope: Lessons from the Howard Dean campaign for the future of Internet politics* (pp. 110–129). Boulder, CO: Paradigm.

Stirland, S. L. (29 October, 2008). Obama's secret weapons: Internet, databases, and psychology. *Wired.* Retrieved 12 January 2009 from http://www.wired.com/threatlevel/2008/10-/obamas-secret-w/

Talbot, D. (8 January 2009). The geeks behind Obama's web strategy. *The Boston Globe.* http://www.boston.com/news/politics/2008/articles/2009/01/08/the_geeks_behind_obamas _web_strategy/?page=1

Talbot, D. (2008). How Obama really did it: The social-networking strategy that took an obscure senator to the doors of the White House. *Technology Review* (September/October).

Teachout, Z. (2007). Something much bigger than a candidate. In Z. Teachout and P. Streeter (Eds.), *Mousepads, shoe leather, and hope: Lessons from the Howard Dean campaign for the future of Internet politics* (pp. 55–73). Boulder, CO: Paradigm.

Thompson, N. (17 June, 2009). Iran: Before you have that Twitter-gasm. *Wired.* Retrieved 17 June 2009 from http://www.wired.com/dangerroom/2009/06/iran-before-you-have-that-twitter-gasm/

Trippi, J. (2004). *The revolution will not be televised: Democracy, the Internet, and the overthrow of everything.* New York: Regan Books.

Tumulty, K. (2003, August 3). The Dean factor. *Time.* Retrieved from*http://www.time.com/ time/magazine/article/0,9171,1101030811–472810,00.html*

Vargas, J. A. (20 November, 2008). Obama raised a half a billion online. *Washington Post.* Retrieved 13 June 2009 from *http://voices.washingtonpost.com/44/2008/11/20/obama_ raised_half_a_billion_on.html*

Wainwright, H., Reyes, O., Berlinguer, M., Dove, F., Fuster i Morrell, M., & Subirats, J. (Eds). (2007). *Networked politics: Rethinking political organisation in the age of movements and networks.* Transnational Institute, Amsterdam.

Wolf, G. (January 2004). How the Internet invented Howard Dean. *Wired.*

Wolffe, R. (2009). *Renegade: The making of a president.* New York: Crown.

CHAPTER SEVEN

Health 2.0 and Managing "Dividual" Care in the Network

MARINA LEVINA

In May 2008 Google unveiled its entry in the burgeoning personal medicine industry: Google Health. Billed as an easy, hassle-free way to gather and organized one's medical records, Google Health's claimed purpose is to put "you" in charge of "your" health information. It purported to rescue individuals from the inconvenience, and sometimes oppression, of medical institutions by putting them in control of their data. This is the latest development in Health 2.0: a growing effort to marry Web 2.0 technology, participatory discourse, and network subjectivity to health care and management. At its core Health 2.0 makes a proposition that access to more medical information leads to greater control over one's health and that, combined, control and information rescues individuals from institutional power. This essay problematizes these assumptions by placing the emergence of health information technology in the larger context of network society. Using Health 2.0 movement as a case study, this essay examines the problematics of individual or "dividual" care in the network. Drawing on Giles Deleuze's theory of control society, it argues that the promises of Health 2.0 movement obscure the rise of network power: a system that functions through production of freedom, participation, and information. Through engagement in affective labor practices participants in the Health 2.0 movement are recoded as "dividuals" whose health is intricately tied to that of the network. Referred to in this essay as the "work of being healthy," continuous work to facilitate information flow and to grow the network is thusly understood by participants as a contribution to their own future health. I argue therefore that individual, or in this case, dividual "health" is articulated through the expansion of the postglobal network.

HEALTH 2.0 AND PARTICIPATION IN THE NETWORK

In February 2009, as part of its stimulus package, the Obama administration allocated $19 billion dollars in incentives to jump-start the adoption of digital medical records (Lohr, 2009). This reflects a larger move to promote wider access and data integration in the health information technology field. Wal-Mart, for example, made a push into the market for electronic health records by developing and distributing cheaper hardware and software technology for physicians in small offices (Lohr, 2009). At the same time, Google Health allows users to keep and send their information as a digital file, easily transmittable to the clinic or accessible online. Google has taken the quest for accessibility seriously, releasing an iPhone application called Health Cloud, which allows users to always have access to their health information. The advertised benefit is the promise of centralized health information at the user's fingertips (Farnham, 2009). The Office of the National Coordinator has also promoted the adoption of health information technology and the development of The Nationwide Health Information Network (NHIN) is billed as a "network of networks."

These health information technology initiatives that further promote interconnectivity between users and clinics are a part of Health 2.0, which is traditionally defined as "the use of social software and light-weight tools to promote collaboration between patients, their caregivers, medical professionals, and other stakeholders in health." (Sarasohn-Kahn, 2008) However, there is more to Health 2.0 then the promotion of information communication technology as an enabler for health care collaborations. While issues of access are important, Health 2.0 sees itself as a movement that stresses community building and patient participation. An expanded definition describes Health 2.0 as a four point system, which emphasizes:

1) Personalized search that looks into the long tail, but cares about the user experience.
2) Communities that capture the accumulated knowledge of patients and caregivers; and clinicians—and explain it to the world,
3) Intelligent tools for content delivery—and transactions &
4) Better integration of data with content. All with the result of patients increasingly guiding their own care (Holt 2009).

Health 2.0 movement also sees improvements in health information technology as a way of stressing wider health care reform and furthering advances and competition in the health care industry:

> New concept of health care wherein all the constituents (patients, physicians, providers, and payers) focus on health care value (outcomes/price) and use competition at the medical condition level over the full cycle of care as the catalyst for improving the safety, efficiency, and quality of health care (Shreeve, Crossover Healthcare Blog).

Finally, and most importantly to this discussion, Health 2.0 movement positions itself as a participatory process, one through which users of health information technology are reconstituted as responsible and active patient-citizens as evidenced by these definitions:

> Health 2.0 defines the combination of health data and health information with (patient) experience through the use of ICT, enabling the citizen to become an active and responsible partner in his/her own health and care pathway (Boss etc., 2008).

Or

> Health 2.0 is participatory healthcare. Enabled by information, software, and community that we collect or create, we the patients can be effective partners in our own healthcare, and we the people can participate in reshaping the health system itself (Eytan, 2008).

This particular discourse of Health 2.0, taken up by numerous advocates, juxtaposes active patient-citizen participation to institutional practices that ail health care system. For example, Scott Shreeve, a prominent physician advocate of the Health 2.0 movement, argues "the traditional paternalism (structural, cultural, regulatory, and political) inherent in medicine is giving way to the participatory nature of Health 2.0." (Shreve, 2009) These narratives hail the revolutionary potential of patient participation, advocate for an overhaul of the traditional care delivery system, and position the Health 2.0 network as an antidote to the institutional medical system:

> Health 2.0 has already changed the landscape of health by delivering tools and technology that empowers patient communities, results in connected physicians, forces transparency to the system, and restores the patient to the center of the health experience. However, much of this has happened at the margins, outside the traditionally paternalistic medical-industrial system. While this has populist and even revolutionary appeal, the quest for far broader adoption of these concepts must penetrate deeper into the underbelly, into the very heart of the plumbing, to attack the calcified hairball where a thousand health revolutions have died before (Shreeve, 2009).

This sentiment is echoed throughout Health 2.0 discourse. Online communities, such as PatientsLikeMe.com, connect patients, provide a virtual space for support groups, but also mine patient data to affect medical research and trials. These sites, as well as access engines such as Google Health, fully embrace and use participatory discourse of the Health 2.0 movement. And their supporters argue that health information technology and social media tools have already had potentially radical effects on the health care industry in general. For example, Shaw (2009) argues that we are witnessing a healthcare reformation equivalent of the 16th century religious Reformation when the printing press and radical thinking took the proprietary interpretation of the Bible away from the Catholic Church. She writes, "traditional paternalistic relationships between patients and doctors are being undermined in much the same way as the religious Reformation of the 16th century empowered the laity and threatened the 1,000-year-old hierarchy of the Catholic church in Europe. The Reformation had irreversible consequences for Western society; the implications of the health -care reformation could also be profound….In our age, the 'bible' is medical information, the technology is the internet, and the priests are the medical profession. The Internet has brought the canon of medical knowledge—previously accessible only in expensive textbooks, subscription journals, and libraries—into the hands and homes of ordinary people." (p. b1080)

And in a *New York Times* article titled "Logging on for a Second or Third Opinion" Dr. Ted Eytan, medical director for delivery systems operations improvement at the Permanente Federation, opines "patients aren't learning from Web sites—they're learning from each other." The shift is nothing less than democratization of health care," he went on, adding, "Now you can become a national expert in your bedroom." (Schwartz, 2008) Here expertise is redefined not through institutional medical training, but rather as access and participation in web-based communities or networks where information is shared amongst many interested parties and individuals.

The shift towards embracing what California HealthCare Foundation's report termed "the wisdom of patients" positions health information technologies within a larger school of thought that emphasizes social media-based wisdom of aggregates, or crowds. Clay Shirky, an author of a well-received book *Here Come Everybody: The Power of Organizing without Organizations,* argued at a 2008 Health 2.0 Conference that "patients in aggregate behave very differently than when solo…what you do when you get a bad diagnosis—you fire up Google, find out who has what you have, and then talk to them. That ability, for patients to pool their resources, is a massive change to the health industry." (Davis, 2008) And the afformentioned "wisdom of patients" posited that collective wisdom of patients, aggregated through social media technologies, can yield knowledge beyond any single patient or doctor. Its authors argued that

Health 2.0 is the result of these trends in accumulation and sharing of collective wisdom; "a new movement that challenges the notion that health care happens only between a single patient and doctor in an exam room." (Sarasohn-Kahn, 2008) Furthermore the report stated that that the characteristics inherent in the technologies that shape social media will generate better, more useful knowledge. Therefore the wisdom of crowds challenges the necessity or dependence on a single expert opinion. The medium may not be the message, but it is health. And information collected through Health 2.0 practices is juxtaposed to knowledge generated through medical institutions. In the report, Sarasohn-Kahn (2008) writes:

> When patients managing the same chronic condition share observations with each other, their collective wisdom can yield clinical insights well beyond the understanding of any single patient or physician. Similarly, when physicians share information with each other online, the results go well beyond the doctor's lounge—the traditional locale for exchanging clinical experiences and insights.

A promotional video for the 2008 Health 2.0 Conference also advocates for collective participation and information aggregation. A retelling of the history of medicine from Ancient China to Web 2.0, the video welcomes us to Health 2.0 and states that "Health is Information Technology; Health is US. Welcome to Health 2.0!" Using, among other things, shots of a terminal cancer patient blogging at CarePages.com from his hospital bed, this video echoes the sentiment of other Health 2.0 NGOs, physicians, and advocates. It is the claim that the Health 2.0 movement is, at its core, a solution to institutional power inherent in the current medical system. Health 2.0 discourse presupposes that with unrestricted access to information, ability to provide online feedback on physicians or treatments, together with social media, participatory patient and physicians practices will ultimately lead to liberation from hierarchical, and often arbitrary, structure of the current health care system. Therefore, the video's statement that health is information technology is a quite literal claim that access to health information technology and organization through social networking tools will not only liberate us, but will, in fact, make us healthy. Moreover, I argue that at stake here is an introduction of a new, post-network, care of self; one that irrevocably ties the health of the individual to that of the network. Whereas in the nineteenth and twentieth centuries citizen bodies were discursively tied to that of the nation-state, in the postglobal network society, citizen bodies serve as stand-ins for the network itself (Levina 2009). In her Health 2.0 report, Sarasohn-Kahn (2008) argues that one of the more important reasons for aggregating health data is preserving and improving the health of the network—what she terms "the concept of intelligent networks." She states that because of the "wisdom of crowds" phenomenon, networks get better as more people use them. The rest of the chapter puts Health 2.0 in the context of network power and subjectivity while arguing how, through affective labor practices, the health of the self is reconfigured in terms of the health of the network.

NETWORK POWER, SUBJECTIVITY, AND HEALTH 2.0

As argued above, discursive positioning of Health 2.0 movement as a form of resistance to hierarchical power of medical institutions often knowingly establishes its practices—such as Google Health, PatientsLikeMe, and CarePages—as outside of functions of power and capital. These claims are common amongst advocates—for instance, Clay Shirky stated during a keynote address at the 2008 Health 2.0 conference that institutions are trying to prevent Health 2.0 from happening. Such discursive moves are sometimes political and more often are rhetorically elusive, but they illustrate an important tension in the postglobal network society—between power/knowledge and control/information—or between power that acts directly on individual bodies through discursive production of knowledge and more diffuse functioning of control, which operates through constant collection and dissemination of information. In Health 2.0 discourse, however, control/information is seen as an antidote to power/knowledge. This ignores how control/information is grounded in different modes of power operations—that of network power.

A proper theorizing of the Health 2.0 movement then needs to be grounded within the rise of a network society and, with it, a system of power relations necessitated by the rise of globalization and the emergence of information technology (Castells 2000, Hardt and Negri, 2000). Castells (2000) argued that network society is characterized by the pre-eminence of social morphology over social action; a logic that privileges network form, expansion, and information flows over any particular social interest—a prioritization of the power of flows over the flows of power. As a non-linear power relation, which operates through decentralized relations of sociability, network power operates through regulations of standards as opposed to the enforcement of a sovereign will. This does not mean that network power is democratic—as David Singh Grewal (2008) argues, "in this case, aggregate outcomes emerge not from an act of collective decision-making, but through the accumulation of decentralized, individual decisions that, taken together, nonetheless conduce to a circumstances that affects the entire group," (p.9)—but rather that it is a diffuse system of control and regulation operating through a multitude of nodes (Grewal 2008) that defines network power as a complex system of coordination and expansion:

First, that coordinating standards are more valuable when greater numbers of people use them, and second, that this dynamic—which I describe as a form of power—can lead to the progressive elimination of the alternatives over which otherwise free choice can effectively be exercised....When these ideas are considered together, the central premise of network power is that the benefits that come from using one standard rather than another increase with the number of users, such that dominant standards can edge out rival ones.

Health information technologies at the heart of Health 2.0 have long been a subject of debate about standardization of information flows. The debate over proper keeping and dissemination of electronic health records is a direct result of disagreements over what counts as an effective patient care, who should be establishing standards for sharing of patient information, and how such dissemination best be implemented. Diamond and Shirky (2008), for instance, argue that one of the dilemmas facing proliferation of health information technologies is the establishment of functioning standards according to which information can be effectively moved across the network. They argue that such standards cannot be simply enforced, but rather must evolve out of what is best for the network itself. They write:

> It seems tautological, but standards aren't really standard unless they are widely adopted, and this step can't be easily mandated. You can't "make" a standard any more than you can "make" friends; people become friends over long association, and so it is with standards. The way something becomes a standard is for it to become standard—which is to say, for it to become the normal case in the field, not merely in the lab or a conference demonstration....At this early stage of evolution in health IT, standards for moving data across the network are more important than standards expressing the content of that data. Put another way, it is better to share important but uncodified information between Doctor A and Doctor B so that an informed clinical decision can be made, than to have perfectly formatted data that never leaves Doctor A's office (p. 385–387).

This example illustrates how Health 2.0 functions as a constitutively social process of network power. Network power operates through decentralized relations of sociability, and as such it is always relational, always circumstantial, and always mutable. It also encourages relations of sociability in order to facilitate expansion. As Michael Hardt and Antonio Negri argue, "network power must be distinguished from other purely expansionist and imperialist forms of expansion. The fundamental difference is that the expansiveness of the immanent concept of sovereignty is inclusive, not exclusive. In other words, when it expands, this new sovereignty does not annex or destroy the other powers it faces but on the contrary opens itself to them, including them in the network." (2000, p. 166) The power of the network is in its continuous and constant growth and openness to divergence and difference (Terranova, 2004). This does not make the exercises of power benign; indeed network power operates through incorporation of dividend elements. Nothing can or should be outside of the network (Galloway and Thacker, 2007). And as such network power is embedded in what has been called a larger "control society."

In his essay "The Postscript on Control Societies," Giles Deleuze (1995) argues that control societies are taking over disciplinary societies. Whereas disciplines operated through institutional confinement that aimed to mold individual bodies, controls are modulations changing from one moment to the next. This is a transition from power that acts on the body, to "the ultrarapid forms of apparently free-floating control." (p. 178) Deleuze writes, "control is short-term and rapidly shifting, but at the same time continuous and unbounded, whereas discipline was long-term, infinite, and discontinuous. A man is no longer a man confined but a man in debt." (p. 181) Control societies can be best understood in terms of networks, where individuals, or "dividuals" as Deleuze calls those living in control societies, are never done, never finish anything—but are continuously moved from one node to another. As Galloway and Thacker (2007) write:

> Control in the networks operates less through the exception of individuals, groups, or institutions and more through the exceptional quality of networks or of their topologies. What matters, then, is less the character of the individual nodes than the topological space within which and through which they operate as nodes. To be a node is not solely a casual affair; it is not to "do" this or to "do" that. To be a node is to exist inseparably from a set of possibilities and parameters—to function within a topology of control (p. 40).

I argue that the Health 2.0 movement through health information technologies and social media discursively reconfigures bodies and identities of its users and participants as dividual subjectivities. For example, Sarasohn-Kahn (2008) argues for development of a service that will enable faster inauguration of various subjectivities into health networks. She writes, "social networks in health are proliferating so rapidly that there is a need for services that "knit" communities together to enable health consumers to move seamlessly and efficiently through the networks without having to be a member of all the groups that pertain to their illness or interest." (p. 15) The efficient move of health consumers through the networks requires its users to think of themselves in terms of information most useful to presentation of their subjectivities. This also indicates that the flow of users in the network is tied to the flow of information.

Deleuze argues that the function of control is to collect, direct, and distribute information. In fact, control societies intrinsically rely on information technologies and computers. Information gives control to its material existence; it's what makes control matter. Alexander Galloway and Eugene Thacker (2007) argue that protocol—a horizontal, distributed control apparatus that guides formations of networks—functions in computer and biological networks when it directs flow of information. In that sense, "information is the concept that enables a wide range of networks—computational, biological, economic, political—to be networks. Information is a key commodity in the organizational logic of protocological control." (p. 57) Generating information gives networks capacity to grow, to regulate, and to circulate. This is the underlying logic, or protocol, of the network. This is also the protocol of social networks embraced by Health 2.0. Sarasohn-Kahn (2008) writes, "the more participants there are in a social network—the foundation of Health 2.0—the more value they create. This is the phenomenon of positive network effects." (p. 5) The value that participants create is to generate more information in the network.

Information flows in the network are not inconsequential; they alter topologies, relationships, and identities. Tiziana Terranova (2004) adds that "the rise of the concept of information has contributed to the development of new techniques for collecting and storing information that have simultaneously attacked and reinforced the macroscopic moulds of identity." (p. 34) Unlike knowledge, information is always in-flux. Knowledge is reified in institutions, while information is continuously changing and continuously in need of control. Without institutions to give it form, information flows through the network—difficult to fix, difficult to manage, difficult to control. Whereas knowledge generates meaning, one could argue that the more information there is, the less meaning there is (Terranova, 2004). Therefore, the constant movement of information in networks encourages volatile spaces, random relationships, and in-flux identities. In the control society, you are your information. As Deleuze points out "the digital language of control is made up of codes indicating whether access to some information should be allowed or denied. We are no longer dealing with a duality of mass and individual. Individuals become "dividuals," and masses become samples, data, markets, or 'banks' (p. 180). Identity constituted by information is identity in-flux. It can always be changed and altered. More importantly it can only be understood in the context of other data. Therefore, in the control society, dividuals can understand themselves only in terms of relationship to others in the network. I argue that Health 2.0 is a product of control societies and as such it expatiates the functioning of information systems. It is dedicated to the collection, distribution, and circulation of information about life itself. Moreover, Health 2.0 dividuals are a fragmented and fractured entity—a subject for "database" information searches, an entity to be classified, and categorized. As Terranova (2004) argues, "the cultural politics of information does not address so much the threat of 'disembodiment,' or the disappearance of the body, but its microdissection and modulation, as it is split and decomposed into segments of variable and adjustable sizes (race, gender, sexual preferences; but also income, demographics, cultural preferences and interests)."(p. 34)

For example, PatientsLikeMe.com—a prominent social network for patients often referenced as a remarkable success story in Health 2.0 literature—presents each patient profile as a series of charts. Divided into sub-sections, such as "Mood Map," "Symptoms," "Counseling/Therapy," "Treatment," and "Weight," a patient is represented through graphs and data sets, which track patient information over the course of days, weeks, months, and years. Patient profiles are a multitude of data—individual experience distilled down to pure information. And while PatientsLikeMe.com provides support forums, the goal of the site is to continuously generate patient data. In fact its mission states:

> PatientsLikeMe is committed to providing a better, more effective way to capture valuable results and share them with patients, healthcare professionals, and industry organizations that are trying to treat the disease....Our goal is to enable people to share information that can improve the lives of patients diagnosed with life-changing diseases. To make this happen, we've created a platform for collecting and sharing real world, outcome-based patient data (patientslikeme.com) and are establishing data-sharing partnerships with doctors, pharmaceutical and medical device companies, research organizations, and non-profits. Contact us if you're interested in working together to achieve our goals (About PatientsLikeMe, 2009).

Here information is generated, distributed, and shared to establish and grow a network. In fact, PatientsLikeMe.com, amongst other Health 2.0 organizations and narratives, draw a direct connection between patient health/disease treatment, generating/sharing information, and network growth. The health of dividual users is directly connected to the overall health of the network. And sharing your health information is considered to be an act of citizenship, precisely because it is good for the network. Much like national health campaigns tied the health of an individual citizen to that of a nation-state, Health 2.0 movement ties the health of a dividual citizen to that of a network (Levina 2009). The testimonial page of PatientsLikeMe feature this anonymous comment:

> I urge all to go to PatientsLikeMe! Each of us working together can one day see a cure come to us. The PatientsLikeMe project does exactly that, it documents and assembles our individual stories in such a way as to provide compelling data to help direct and focus researchers and advocates. I urge all in the group to go to PatientsLikeMe.com and submit your information! (Testimonials, 2009).

This call echoes others in the Health 2.0 movement. As a discourse they reconfigure allegiance of a citizen away from a particular nation-state and toward a postglobal network. Moreover, as testimonial above indicates, these citizenship actions require subjectification of bodies and identities to the goals of the network. They also require lots of work on the part of participants. The time and effort dedicated to the upkeep of user profiles on PatientsLikeMe.com seem extraordinary. As a result, participation in the network does not only expand network power, but also reconfigures labor practices in terms of affects. This allows for the Health 2.0 movement to position work performed as outside of structures of capital. If work is connected to future health then, in the network, work is healthy.

CONCLUSION: THE WORK OF BEING HEALTHY—AFFECTIVE LABOR OF HEALTH 2.0

Over the past decade scholars started to address how emergence of globalization and information-based economy have restructured the flows and practices of capital in contemporary society. For example, Deluze wrote that, in control societies, the work of capital is reassigned away from production and towards distribution. And Hardt (1999) has argued that the passage of a paradigm from domination of industry to that of service and information involved a change in the quality and nature of labor practices. He argued that "since the production of services results in no material and durable good, we might define the labor involved in this production as *immaterial labor*—that is, labor that produces an immaterial good, such as a service, knowledge, or communication." (p. 94) Immaterial labor also produces affect, or "the bodily capacity to affect and be affected or the augmentation or diminution of a body's capacity to act, to engage, and to connect." (Clough, 2007, p. 2) Affective labor does not generate direct financial profit, but rather produces a sense of community, esteem, and/or belonging (Gregg, 2009). Most of the recent work on affective labor has placed it within the realm of digital economy. Evidenced, for example, in the numerous fan sites that painstakingly catalogue television shows and movies, affective labor in digital economy has largely involved what Andrejevic (2004) termed the "work of being watched." This also involves various activities that we perform online—all dedicated towards generating the most prized commodity in the network: information. However, no one could claim that any of this work is making us healthy. That is, until the Health 2.0 movement which is directly involved in the work of being healthy.

As Hardt (1999) argues, networks of affective labor produce collective subjectivities that are then directly exploitable by capital. For example, PatientsLikeMe blog introduces the notion of "Network Patient," one whose labor contributes to the growth of research capital and whose health is irrevocably tied to that of the network:

We believe in the concept called "The Network Patient"—an approach that puts patients first by giving you what you need to know when you need to know it, and empowering you to act on that information. As members of PatientsLikeMe, you have chosen to embrace openness and take control of your health. You volunteer your health information, your experiences, your life—all in an effort to improve your care, support others, and move research forward (Piscatelli Scanlon, 2009).

This is a free labor moment where the productive activity of generating information is embraced for its affects, but also exploited by capital activities. A fact not often mentioned is that much of the Health 2.0 movement, including PatientsLikeMe, is founded by for-profit organizations. This reveals a tension at the heart of the Health 2.0 movement—how to claim a revolutionary position outside of institutional power and capital structure, and, at the same time, accrue financial benefits. Usually this is accomplished through introducing openness as an antidote to capital. This tension is often not easily resolved as evidenced by the following comment section reacting to the blog post above:

I believe if PatientsLikeMe wants to be open it should be clear that it is a for-profit company and that the information collected will be sold on this basis. Therefore the information compiled may not be available to academic and non-profit organizations doing research but cannot afford to pay for your information. They is a particularly valid point in the ALS community since most of the work is being done in these areas and they will not benefit from this if they are unable to pay. What your company is doing is admirable but if you want to promote openness then be upfront that the company is [not] for altruistic purposes (Jeff Lester).

Jeff,
Thanks for the thoughtful message. We endeavor to be completely transparent about being a for-profit company. On our home page for example, at the bottom we have a question that asks, How do we make money? We also have FAQs that state specifically that we are a for profit company and how we sell the data. We do share our information with academic and non-profit institutions for free at times. It often depends on the research goals and the hypothesis being tested. In the ALS community we have completed much research and shared data with organizations fully gratis. Research is at the heart of our company and we will continue to work with organizations to promote accelerated research. Because we are small, there's only so much we can do on our own so collaborations are important. You will continue to see more in the coming months (David S. Williams III).

However, to call the work of these companies simply exploitative or deceptive is to ignore the affective part of Health 2.0 labor. In all truth, health data mining will lead, in all probability, to beneficial medical breakthroughs. Therefore, the work of being healthy performed by users of PatientsLikeMe.com is affective in every sense of the word—it is manifested in its ability to produce communities and esteem, but also to enhance bodily capacity to affect and be affected. Therefore, the work of being healthy needs to be understood as part of a bioeconomy that produces life itself. As Hardt writes "what is created in the networks of affective labor is a form-of-life....Biopolitical production here consists primarily in the labor involved in the creation of life—not the activities of procreation but the creation of life precisely in the production and reproduction of affects....Labor works directly on the affects; it produces subjectivity, it produces society, it produces life. Affective labor, in this sense, is ontological—it reveals living labor constituting a form of life and thus demonstrates again the potential of biopoliticial production." (pp. 98-99)

I argue here that the work of being healthy produces network subjectivity, which understands the care of self as the care of the network. These dividual subjectivities function based on the principle of network power—rule through decentralized relations of sociability induced by affective labor practices. Therefore, in agreement with Terranova (2000), I argue that we need to examine the new logic of value produced by Health 2.0 discourse, one that connects individual health to that of the network; the logic that insists on the value of the constant labor of participation. To volunteer one's life is to subject oneself to openness, without which the network ceases to exist. Perhaps what the Health 2.0 movement demonstrates are the consequences of life in the (post)global network, where our health is that of the network. Whether this will be a symbiotic relationship is yet to be determined.

REFERENCES

About PatientsLikeMe (2009). Retrieved June 4, 2009, from http://www.patientslikeme.com/about

Andrejevic, M. (2004). *Reality TV: The work of being watched*. Lanham, Md.: Rowman & Littlefield Publishers.

Bos, L., etc. (2008). Patient 2.0 Empowerment. Patient 2.0 Empowerment Proceedings of the 2008 International Conference on Semantic Web & Web Services SWWS08, Hamid R. Arabnia, Andy Marsh (eds), pp. 164–167.

Castells, M. (2000). *The rise of the network society (2nd ed.)*. Oxford; Malden MA: Blackwell Publishers.

Clough, P. (2007). *The affective turn : theorizing the social*. Durham: Duke University Press.

Davis, L. (2008, October 22). Health 2.0 and the new economics of aggregation. ReadWriteWeb. Retrieved April 22, 2009, from http://www.readwriteweb.com/archives/health_20_economics_of_aggregation.php.

Deleuze, G. (1995). *Negotiations, 1972–1990*. New York: Columbia University Press.

Farnham, K. (2009, March 30). Google Health: Now on Your iPhone. Programmable Web. Retrieved April 10, 2009, from http://blog.programmableweb.com/2009/03/30/google-health-now-on-your-iphone/

Eytan, T. (2008, June 13). The Health 2.0 Definition : Not just the Latest, The Greatest! E-Health. Patient Empowerment. Retrieved August 10, 2009, from http://www.tedeytan.com/2008/06/13/1089.

Galloway, A. (2007). *The exploit : A theory of networks*. Minneapolis: University of Minnesota Press.

Gregg, M. (2009). Learning to (Love) Labour: Production Cultures and the Affective Turn. *Communication and Critical Cultural Studies, 6*(2), 209–214.

Grewal, D. (2008). *Network power : the social dynamics of globalization.* New Haven: Yale University Press.

Hardt, M. (1999). Dossier: Scattered Speculations on Value-Affective Labor. *Boundary 2., 26*(2), 89.

Hardt, M. (2000). *Empire.* Cambridge Mass.: Harvard University Press.

Holt, M. (2009, February 3). Health 2.0: User-Generated Healthcare. Retrieved June 10, 2009, from http://www.ncvhs.hhs.gov/090226p3.pdf.

Levina, M. (2009). Regulation and Discipline in the Genomic Age: A Consideration of Differences between Genetic Engineering and Genomics. In S. Binkley & J. Capetillo (Eds.), *A Foucault for the 21st century: Governmentality, biopolitics and discipline in the new millennium* (308–319). Newcastle: Cambridge Scholars Publishing.

Lester, J. PatientsLike Me Blog. Retrieved June 14, 2009 from http://blog.patientslikeme.com/2009/06/11/sharing-is-a-right-as-well/#comments

Lohr, S. (2009, March 11). Wal-Mart Plans to Market Digital Health Records System. The New York Times. Retrieved April 23, 2009, from http://www.nytimes.com/2009/03/11/business/11record.html?_r=2.

Piscatelli Scanlon, L. (2009, June 11). Sharing Is a Right as Well. PatientsLikeMe Blog. Retrieved June 14, 2009, from http://blog.patientslikeme.com/tag/the-network-patient/

Sarasohn-Kahn, J. (2008). The Wisdom of Patients: Health Care Meets Online Social Media. ihealth reports (pp. 1–24). California HealthCare Foundation. Retrieved May 12, 2009, from http://www.chcf.org/topics/chronicdisease/index.cfm?itemID=133631

Schwartz, J. (2008, September 30). Logging On for a Second (or Third) Opinion. The New York Times. Retrieved October 1, 2008, from http://www.nytimes.com/2008/09/30/health/30online.html.

Shaw, J. (2009). A Reformation for Our Time. BMJ, 338, b1080.

Shreeve, S. Crossover Healthcare Blog. Retrieved April 22, 2009, from http://blog.crossover-health.com/

Shreeve, S. (2009, April 2009). Building Health 2.0 Into the Delivery System. Retrieved May 1, 2009, from http://blog.crossoverhealth.com/

Terranova, T. (2000). Free Labor: Producing Culture for the Digital Economy. *Social Text, 18*(2), 33–58.

Terranova, T. (2004). *Network culture : politics for the information age.* London; Ann Arbor MI: Pluto Press.

Testimonials (2009). Retrieved June 4, 2009, from http://www.patientslikeme.com/about/testimonials

Williams, D. PatientsLike Me Blog. Retrieved June 14, 2009 from http://blog.patientslikeme.com/2009/06/11/sharing-is-a-right-as-well/#comment

CHAPTER EIGHT

Sport in the Wires

Abstraction, Integration, Efficiency

SEAN SMITH

INTRODUCTION

When one is to consider sport in the age of the post-global network, perhaps the obvious locus of investigation is how new forms of computer-mediated communication have changed the fan experience and the business of sports spectatorship. Indeed, throughout the past century sport has been one of the proving grounds, so to speak, for the introduction of new media technologies to the consumer market, in part due to its ability to engage individuals in ever-contingent notions of community (Raney & Bryant, 2006; Lowes, 1999; Wenner, 1989). Since the dawn of electricity-based communication with the telegraph, sport has provided an early form of content and, indeed, actively shaped the structural forms of the medium. But behind the scenes of the burgeoning sports spectacle, the numerical data that was driving telegraph communication was simultaneously being used to rationalize performance on the field of play (Puerzer, 2002). So if we are to focus strictly on the easy and most readily apparent shifts in the relationship between sport and the network today, taken from the perspective of spectatorship or consumption (for example with sports blogs, videogames or cellphone applications), we risk missing changes in the nature of professional sport production that are equally bound up in the assemblies of local area network, regional communication hub and transcontinental data conduit, and which perhaps have more significant conclusions beyond sport for the future of labour practice.

This essay examines the post-global network from the perspective of a media ecosystem, building upon foundational work in media ecology by Marshall McLuhan (1964), Neil Postman (1992), and Paul Levinson (1997), as well as more contemporary work on the relationship between network society and political economy by Manuel Castells (2001; 2000). In this approach, the material forms of communication themselves play an important formative role in shaping human perception and the conditions of possibility for society. McLuhan (1964) distinguished between pre-literate, literate, and post-literate societies, which he classified based on their dominant mode(s) of communication. Pre-literate societies communicate primarily by the spoken word; literate societies emerged from the introduction of the phonetic alphabet and the Gutenberg press and communicate predominantly through book form; and post-literate societies are those that are characterized by the electric communications technologies of telegraph, telephone, radio, television, personal computer, satellite, and so forth. McLuhan's hypothesis was that post-literate societies—that is, those who live in the electric age of communications, such as ourselves—would very much resemble pre-literate societies in the way that they acted, both as individuals and as a collective.

With particular resonance for this essay, McLuhan argued that electricity and digitality fundamentally changed the modes of social organization in contemporary societies. For McLuhan, the reliance on auditory and tactile forms of perception for the pre-literate society had consequences in the way that space and time were perceived: the former as a resonating sphere and the latter as a circular process. With the introduction of linearity in alphabetic writing, accelerated by the distribution potential of the printing press, the perception of space and time similarly began to shift: the former became a geometric container in which perspectival vision extends and vanishes at a point on the horizon, and the latter became an unfolding of a linear process that begat the cause and effect thought of the Enlightenment.

An ecosystems approach suggests that flows—of people, goods, information, money—are key to understanding the nature of production and consumption in contemporary society. Yet these flows are still constituted and constrained by material forms and structures. Castells (2000) identifies three such layers of material structure that constitute what he refers to as the "space of flows." The first is the layer of electronic exchange, the combination of hardware, software and communication protocol that allows for the most basic existence of the network society. The second material layer is geospatial, which channels the essential placeness of the network into more or less well-defined aggregations, congregations and access points in physical space. The third material layer of flows is constituted by those homogenous spaces in which the global political and financial elites conduct their everyday work and leisure.

Professional sport cuts across each of these material layers in distinct, though interrelated ways. Since the production of sporting contests creates a great deal of content in the form of statistics and images that are transmitted across the first material layer of electronic exchange, professional sport franchises have become attractive acquisitions for vertically-integrated media and communication corporations. While this data is synergized across a wide variety of platforms, it most certainly condenses in cultural zones of attraction such as the home living room, sports bar, or casino sports betting book. As the site of simultaneous production and consumption, the stadium is the most significant of these zones, but its luxury boxes and VIP clubs also constitute the space in which global elites may combine leisure with the communicative and relationship-building elements of work—and it may also be where the ownership and management of professional sports teams can share ideas with those in other business sectors.

Sport in the wires. The phrase seems an oxymoron, for sport is fundamentally about moving bodies, while the discourse structuring the network was that of a communication architecture using electricity and information to connect static "terminals" and the docile users sitting in front of their screens (DeLanda, 1998). But as the interfaces of computer input and output have become far more dynamic, the contemporary post-global network has similarly become exceedingly sophisticated at adjudicating bodies in motion. For this reason, sport in the wires should be understood as an investigation at the threshold, the liminal space where moving athletic body "touches" or is inscribed upon the archives of material communication technology and becomes immaterial flow of data, the latter of which has become essential to contemporary professional sport in the drive for production efficiencies on the field of play.

In the late 1970s an amateur statistician named Bill James recognized that the existing metrics used to measure performance in baseball did not accurately reflect true production value and he began to develop new metrics of his own. Nearly two decades later, Oakland Athletics general manager Billy Beane recognized the potential of James' work to discover undervalued talent in the highly competitive professional baseball labour market. Recognizing that past performance was a better predictor of future performance than the theretofore privileged knowledge of baseball scouts, Beane began scouring the internet for the statistics of college baseball players and submitting them to econometric analyses based on James' pioneering work, going so far as to select a player sight unseen in the 2002 Major League Baseball (MLB) entry draft (Lewis, 2003). Beane's use of the predictive potential in statistical modeling to revalue prospects and assemble a successful baseball team despite Oakland's small-market revenue disadvantage was enabled by the proliferation of athletic department web sites that together featured the game statistics of nearly every college player in the nation. It was thus a democratic shift in the ability to publish on the network that changed baseball's economics from a production orientation to one based in simulation (Baudrillard, 1983), statistical models anticipating the baseball production process in advance.

While Beane's competitors in Major League Baseball were eventually forced to adopt similar analyses within their own organizations, enterprising managers in other sports also began to investigate how econometric techniques might achieve similar competitive advantage. But every sport is unique in its material and semiotic conditions of possibility. For example, soccer is a sport with few goals scored, few other native statistics and few striations of the playing surface from which metadata may be created or extracted (Deleuze & Guattari, 1987). To compensate for these differences, video imaging devices were added to the network apparatus in efforts to rationalize performance in soccer. In this case, a system called ProZone provides analytics that measure productivity during a match using multiple cameras that surround the stadium to capture all player movement from a variety of angles. These multiple feeds are processed in real-time by a proprietary software package that triangulates, traces and tracks each athlete as a unique data-object on the pitch, all without the assistance of any sort of motion capture marker on the athlete's body. Once tracked, managers may analyze variables such as work output and pass efficiency, share the information across the organization, and sell it to sport media companies.

This suggests that sport in the wires should also be understood as an investigation located at the thresholds between various communication technologies, given the understanding that the post-global network is a media ecosystem combining speech, written word and image in forms either analog or digital. These flows of communication may belong to particularly focused or diffused geographies, and may be more or less freely accessible to those interested in the information. Stocks of data, in the form of archived game statistics and multimedia images, also have important value in this media ecosystem: "The storage of information may be as valuable as its transmission, and the archive is a vector through time just as telesthesia is a vector through space." (Wark, 2004, #318) While this essay is primarily concerned with those communication forms enabled by the internet and associated protocols, it should be stressed at the outset that the revolution in sporting econometrics described here as sport in the wires would not have accelerated in the same fashion (nor perhaps even been possible) were it not for the complex assemblies of material and semiotic forms at the media ecosystem level. And it certainly should not be perceived in a purely positive sense. While Parikka (2007) reminds us that the media ecosystem is susceptible to biological processes such as decay and contagion, Shaviro flatly criticizes the metaphor altogether, warning that the natural evolution of the media ecosystem is decisively bound with politics: "Such is the soft fascism of the corporate network: it reconciles the conflicting imperatives of aggressive predation on one hand, and unquestioning obedience and conformity on the other." (2003, p. 4)

As such, we shall consider professional sport production and consumption as part of a media ecosystem that weaves organic and technical elements together, keeping in mind the consequences in political economy this implies. The following study explores how professional sport and network technologies co-emerged over the past century to create the conditions under which econometric principles could be applied to the manufacture of professional sport contests. In the age of the post-global network, these drives for abstraction, integration and efficiency has spread beyond its original forms in baseball to move into other professional sports, mutating the tools of analysis in the process. In turn, advanced sporting econometrics have become packaged for spectator consumption through various media platforms, normalizing these three drives across the spectrum of labouring classes in the audience.

SMOOTH AND STRIATED SPORTSCAPES

An important consideration in a media ecosystem approach is that the flows of communication are intimately bound with and actively shaped by the spatial environment in which they exist (Castells, 2000; McLuhan, 1964). In *Discipline and Punish: Birth of the Prison*, Michel Foucault (1977) locates a formula of social governance developed through modernity that organized space, time and movement so as to maximize productive utility and minimize political instability. He locates these "disciplinary" practices across a number of institutions, most notably to administer workers in factories, soldiers in military barracks and convicts in prisons. These disciplinary societies are quite consonant with McLuhan's (1964) media ecology of the literate age in that they are both grounded in discrete spatial partitioning, linear causality and the perspectival gaze or point of view.

Deleuze and Guattari (1987) recognize the essentially mutable nature of these disciplinary spaces, however, and develop the paired concepts of striated and smooth space to deal with the question of moving and circulating bodies. They take the strategy of discipline identified by Foucault as the general model of the striated space of State power. Striated space is the enclosed space of the State apparatus, a coded series of points and gridlines that allows for the objects within to be administered and for flows to be constricted or diverted through this measurement, fulfilling "a need for fixed paths in well-defined directions, which restrict speed, regulate circulation, relativize movement, and measure in detail the relative movements of subjects and objects." (p. 385) In turn, as these codings are archived or recorded they become flows of information that may enter a communication network apparatus.

This rationalization of space and time was a fundamental element of the passage from premodern to modern institutions, and played a central role in the discourse of sportification as well (Guttman, 1978). As the trajectory of this modernization continues apace, sport scholars have leveraged the work of Foucault to likewise understand athletes as objects of disciplinary technology, in which constraints of space, time and movement enable coaches, trainers and the athletes themselves to wrest maximal potential from the competing body by coordinating and channelling this complex of forces towards the most advantageous fulfillment of the sport's rational goals (Markula & Pringle, 2006; Shogan, 1999; Bale, 1994).

For example, the playing area of a baseball diamond is a relatively striated space. A 90-degree arc of a circle, with the batter at the centre point of origin, creates a vector of force in which all balls hit must travel in a particular direction in order to be considered "fair" or legal. The distances between the bases, the distance from pitching mound to home plate, the dimensions of the batter's box: all highly rationalized and standardized. Only the distance to the outfield wall is considered variable under baseball rules, and thus short or long fields provide the differentiation between "hitter's" and "pitcher's" parks. Once hit, the batter moves on a linear progression as if on an assembly line (cf. McLuhan, 1964) through the bases in order to produce a run scored. And as Michael Lewis points out in his book *Moneyball*, runs scored are "the money of baseball, the common denominator of everything that occurred on the field," particularly in the professional context (2003, p. 131). In the process of producing runs, each time one touches a base in transit or catches a moving ball is an opportunity for measurement and archival inscription. The remote gaze of the umpire legislates the proceedings on the field, the remote gaze of the scorer records events on the official scoresheet, and the remote gaze of the team manager optimizes the allocation of player assets during the course of play.

Smooth space, by contrast, is the space of the nomadic war machine (Deleuze & Guattari, 1987). It is deterritorialized space, flowing space, permeable space—a space of vectors that lacks a centre point. In contrast with the striated, smooth space can be considered as the space that exists, or comes into being, *between* the measurable points: "…in the case of the striated, the line is between two points, while in the smooth, the point is between two lines…." (Deleuze & Guattari, 1987, p. 480) In other words, smooth spaces do not readily offer opportunities for the extraction of numerical data, for there are no points at which one intersects with a formal measurement apparatus. Passage through this space constitutes a *trajectory* that is experienced affectively and intensively.

We should note that these are ideal types for Deleuze and Guattari. In practice, smooth and are striated spaces tend to exist in admixture or combination, and further are not homogenous over time: smooth spaces become captured for administrative purposes and striated as time unfolds, while striated spaces include the conditions of possibility for new types of smooth space to emerge and flourish. There is another important distinction between the two types of space: striated space is optic while smooth space is haptic (Deleuze & Guattari, 1987). Foucault (1977) already understood the optic nature of disciplinary space and its surveillant gaze, which takes the remote point of view in its adjudication of subjects. Haptic perception, on the other hand, entails sensuous, tactile participation that eliminates the remote perspective of the gaze in favour of close, immersive sensation. As we shall see later in our investigation of the ProZone soccer system, this threshold at which smooth becomes striated may be facilitated by haptic forms of perception that function in the service of the remote perspectival optics characteristic of disciplinary space.

Striated and smooth are thus relative terms, particularly as they exist in the context of sporting spaces and their related media environments. But by understanding how the forces of striation allow for the measurement and archiving of sporting performance, we shall in turn have a better understanding of how network dynamics integrate with sites of sporting production in a media ecosystems sense, and how the materiality of athletic movement becomes the immateriality of statistical data which thereafter reshapes that very movement.

BASEBALL AND ELECTRICITY: A BRIEF HISTORY OF THE PRESENT

In order to understand some of the contemporary changes in the econometric analysis of sporting labour, it is necessary to take a brief historical detour to explore early attempts to archive the game of baseball, transmit it across a burgeoning network of telegraph lines, and rationalize its inputs for monetary gain. The work of Richard Puerzer (2002) sheds light on these three parallel strands in the history of baseball record-keeping, particularly towards understanding the subsequent integration of these archives into baseball production as a problem in industrial engineering and management. In this analysis, baseball is intimately entwined with the historical conditions of capitalist production, which is reflected within the decisions and techniques used both on the field and in the front offices of professional clubs.

As baseball evolved in the late nineteenth-century, there arose a demand for a system of notation to efficiently record baseball performances (Puerzer, 2002). Enter Henry Chadwick, the American sportswriter known as the "Father of Baseball," who grew up in England and was raised on cricket. Despite being initially underwhelmed upon his introduction to baseball after arriving in the United States, Chadwick became very passionate about the game and was instrumental in raising its profile and popularity through his journalistic endeavors (Tygiel, 2000). Though baseball is directly descendent from the game of rounders, Chadwick recognized its earlier genealogical roots in and structural similarities with cricket, and sought to apply the accounting principles found in the scorecard of the latter by developing the boxscore system for baseball (Puerzer, 2002). There were certain important distinctions between the two sports, however, which led to differences in how baseball was archived. The first involved the pace of play: the baseball of industrial America was a much faster-paced game than cricket, designed to be played in a few hours following a typical workday. As such, Chadwick's boxscore system required a shorthand notation to quickly record each discrete event within the game. While both the cricket and baseball systems featured several abbreviations to represent key events, Chadwick's most important innovation involved how the players themselves were archived. In contrast to the cricket scorecard, which used the players' names to record relevant game information, Chadwick's baseball scoring system featured numerical identifiers to mark each athlete's positional location on the field of play and the sequence of events that led to an out being recorded. In keeping with the character of Foucauldian disciplinary space, Chadwick's boxscore system assigned each position a number to code the distribution of bodies in the enclosed space of the baseball diamond.

Thus, a 6–4–3 on the baseball scorecard tells us that a double play occurred in which the shortstop threw to the second baseman, who tagged the base and

Table 1. Baseball position symbols and abbreviations, first developed by Henry Chadwick and shared by nearly all contemporary scorekeeping systems.

Position Number	Position	Position Acronym
1	pitcher	P
2	catcher	C
3	first base	1B
4	second base	2B
5	third base	3B
6	shortstop	SS
7	leftfielder	LF
8	centerfielder	CF
9	rightfielder	RF

threw to the first baseman ahead of the runner, notching two outs in the process. While elegant and compact, there are limitations to this form of record-keeping. A spatiotemporal, embodied fluidity and variability unites the various components of the 6–4-3 assembly that is not captured by the archival process—it is a *low definition* form of information shorthand. Furthermore, it does not necessarily let us know *who* was involved in the 6–4-3 play. If several athletes play the shortstop ("6") position during a game, to whom should the administrative gaze assign statistics in the sport performance archive?

At about this time in the development of baseball we begin to see the introduction of jersey numbers on team uniforms, and soon thereafter, the archiving of baseball statistics by player number. The specific origin of uniform numbers in baseball is a matter of considerable uncertainty, though there appears to have been some cross-pollination of the practice with other team sports played at the cusp of the twentieth century, such as football, hockey, and basketball. James (1988) suggests that the Cincinnati Red Stockings experimented with uniform numbers as early as 1883, but an absence of corroborating testimony casts doubt upon its veracity. The National Baseball Hall of Fame (n.d., Numbers and Names section) asserts in its online exhibit that the Reading Red Roses of the Atlantic League intended to use uniform numbers in 1907, though wanting for photographic documentation it is uncertain whether or not this actually occurred. What is certain is that the exhibit also features a photo from 1909 that shows José Mendez of the Cuban Stars in a uniform with a number emblazoned on the left sleeve. As for professional baseball in the United States, it is known that in 1929 the Cleveland Indians and New York Yankees became the first National League teams to wear numbers on their uniforms for an entire season, and by 1932 the practice had become standard across all of the major leagues (Stang & Harkness, 1996; James, 1988). We see Foucauldian administration at its sporting apex: every athlete in the game under the watchful gaze of the managers and umpires who governed the proceedings, each partitioned into productive spaces on the field, each an easily-identifiable "object of information, never a subject in communication" (Foucault, 1977, p. 200) for which performance was to be accounted, analyzed, and archived.

These administrative solutions in baseball developed in parallel with the rise of Taylorism, as well as the emergent stock exchanges in the United States (Puerzer, 2002). Only a short time after the inaugural communiqué of the Morse telegraph between Washington and Baltimore in 1844, baseball scores were being relayed throughout the burgeoning network of cables around the nation (Tygiel, 2000). At the outset of the sports media ecosystem, we witness the same technology used to power the stock market tickers being used to develop the professional baseball industry. As Marazzi (2009) might suggest, there is an inextricable relationship being developed between capital and linguistic forms: the code that was used to relay baseball information in turn powered large electromagnetic scoreboards in town squares throughout the nation. This gathered crowds for a nearly real-time experience of a remote sporting event—the first of the new networked spaces of flow—and set in motion what has become a multi-billion dollar global sports-media industry (Tygiel, 2000).

"Only in baseball can a team player be a pure individualist first and a team player second, within the rules and spirit of the game," Branch Rickey noted over a half century ago. Rickey, the general manager of the Brooklyn Dodgers, then attempted to rationalize this individuality and incorporate the principles of scientific management into baseball by analyzing the statistics that had emerged from Chadwick's boxscore innovation decades earlier (Puerzer, 2002). But as we shall see in the following section, the structural form of the archival apparatus may also serve to obfuscate attempts by management to improve efficiency on the field of play, creating market opportunities in the process.

THE MONEYBALL REVOLUTION

The business of professional baseball in the United States was changed dramatically in 1975 with the advent of free agency. Prior to that time, baseball had maintained limited labour mobility through its reserve clause, which essentially kept a player indentured to a particular team in perpetuity, thereby keeping salaries artificially low relative to what free market economics would otherwise dictate (Quirk & Fort, 1997). In 1975, Andy Messersmith and Curt Flood successfully fought to overturn baseball's reserve clause in the Supreme Court, ushering in the era of free labour mobility. Baseball salaries skyrocketed almost overnight as players entered the free market and captured a greater share of their produced marginal revenue. This had a number of consequences for the business of baseball, not the least of which were that statistics came to play a greater role in objectively assessing the value of a player, owners needed to find new revenue streams or intensify old ones to recoup some of their lost profits, and the financial risk associated with signing the wrong player magnified dramatically.

With regard to this first consequence, free agency created a stress to the system of archiving athletic performance: no longer was the team the fundamental unit of identity in professional baseball, as the athlete now began to assert a greater degree of market power. While statistics became even more relevant to the economic system of baseball and assumed greater significance in creating the sportocratic vitae of this newly mobile labour, an increase in labour mobility also meant that a greater number of athletes would end up wearing a particular jersey number for a given team over a period of time, which challenged its usefulness as an administrative solution, particularly in creating and maintaining archival data over longer periods of time. A new solution was required.

It was at about this time that industrial America was migrating to a new type of production technology, the microcomputer. These digital calculating machines vastly simplified the process of drawing up tables, classifying individuals, and creating records, a problem that had been made increasingly difficult in baseball with the newly emerging labour mobility. As the first significant architectural form of post-industrial society was born—the relational database—one could now easily create and maintain tables of classification as well as automatically link data from one table to the next (Castells, 2001; Castells, 2000; Levinson, 1997). Hence, the primary key of the database became a type of meta-identifier to the archives of the professional sport network—while the jersey number was still used for the real-time indexing of data to the baseball scorecard, the database primary key assumed hierarchical prominence in baseball's administrative numeration and electronic sharing of information.

As for the latter two consequences of free agency outlined earlier, teams in large urban markets such as New York, Boston, Chicago and Los Angeles were through a combination of population density, productive capacity and historical tradition better able to navigate the economic consequences of increasingly mobile baseball labour. More tickets could be sold at the gate for higher prices and more advertising revenues could be earned on local television in these larger markets. With these financial resources at hand, the payrolls of the richest teams soon began to dwarf those of the poorer ones and they began to purchase as many star players as possible on the free agent market.

While some of these poorer franchises lamented the economic inequities built into the system of professional baseball and sought structural changes at the league governance level, a few others had recognized the need to strategically change the parameters of competition in order to succeed. Notable among this last group was the Oakland Athletics franchise, which had among the lowest payrolls in all of baseball yet somehow managed to compete on the field and in the standings with the richest teams in the late 1990s and early 2000s. As documented in Michael Lewis' *Moneyball* (2003), their success resulted from a perfect storm in which econometric principles were brought to bear on the sport of baseball and the financial valuation of baseball talent.

The story begins in the 1970s, when a baseball enthusiast and amateur statistician named Bill James recognized that many of the existing metrics for baseball (such as batting average and stolen bases) did not accurately measure true production value and began to develop new metrics of his own. He self-published his ideas along with whimsical references to the arcane minutiae of baseball that he came across during the course of his research to a growing network of like-minded baseball statistics enthusiasts through a mail order print newsletter promoted with a one-inch ad in the depths of *The Sporting News*. With its roots in the application of scientific management principles to baseball a century earlier, James was concerned with the objective analysis of baseball performance, usually seeking to optimize production (ie. runs scored) through the positive analysis of baseball statistics. The emerging field of study was dubbed *sabermetrics*, a neologism of econometrics derived from the acronym SABR, for the Society of American Baseball Research.

Two decades later, Oakland Athletics general manager Billy Beane read James' work and realized its potential to discover undervalued talent in the highly competitive professional baseball labour market. The tipping point in professional baseball came when the cost of the athletic factors of production escalated greatly after the advent of free agency. When coupled with baseball's longtime anti-trust exemption, this created the conditions in which, as Lewis states, "what used to be a thousand-dollar mistake was rapidly becoming a million-dollar one." (2003, p. 4) Beane recognized in these new metrics the beginnings of an econometric analysis that would surreptitiously change the parameters of competition by developing pricing models for baseball run production. Applying econometric analysis to the professional baseball archive could provide insights to rationalize the market for baseball labour and allow management to achieve increased cost efficiencies in the pursuit of run production (Lewis, 2003). Beane then used these models to re-create his entire organization from the major league team right down through all the levels of the minor league farm system. For the next several years, Beane's complete overhaul of the Oakland organization allowed the club to compete successfully with teams having larger payrolls, since their pricing models provided competitive advantage within the closed market for baseball labour.

For Beane to be able to reshape his organization so thoroughly and quickly required information and the vectors of transmission to process and communicate this information. What types of information were these baseball players producing? As Lewis (2003) points out, the information of value to baseball's talent scouts and decision makers had been largely qualitative in nature before Beane's transformation of the Oakland Athletics. As Foucault might suggest, in examination each professional baseball hopeful became "the object of information, never a subject in communication." (1977, p. 200) But the character of this objectification is key to understanding the athlete as a factor of production in creating baseball runs, and ultimately, winning outcomes in uncertain game situations. Baseball scouts had far less use for player statistics than they did actually seeing an athlete in the flesh. As the scouts were interested in maintaining their "expert" status within the baseball organization after years of gaining hard-won conventional wisdom, Lewis suggests, this implied hierarchical power structures and personal biases in the system of evaluation. Beane and the Oakland Athletics used advanced statistical methods to internally eliminate these biases and exploit systemic inefficiencies in the pricing market for baseball players.

Within baseball, this marked a decisive departure from the era of Foucauldian disciplinary production, which featured the expert knowledge of a select group of gatekeepers (the scouts); the confessional aspects of baseball prospects undergoing tests of skill and athleticism; and the discursive elements of what constituted a good baseball player in language, physical appearance and performance statistics. Only the numerical aspects of the latter have grown in importance in this new era of the statistical model or code. This echoes what Baudrillard (1983) suggested was the "ephemeral" character of serial production—which we would understand in baseball as the techniques required for the accumulation of runs produced. In the era of simulation it is the *model* that stands as the foundation of political economy, and it is "the very possibility of industrial production that we should look for in the genesis of the code and the simulacra." (Baudrillard, 1983, p. 101) Beane's transformation of the Oakland Athletics through sabermetric analysis stands as Baudrillardian simulation writ large.

Inefficiencies may exist in any highly competitive market, but once discovered they will not remain unexploited for long. The impact of Beane's sabermetric revolution in Major League Baseball has been stunning and swift. J.P. Ricciardi and Paul DePodesta, key management figures in Oakland, are now the general managers of Toronto Blue Jays and Los Angeles Dodgers, respectively. Theo Epstein, a devotee of sabermetric philosophy, landed the general manager job of the large-market Boston Red Sox and brought the franchise two World Series titles after an 86-year drought. With the popular reception of *Moneyball* and the subsequent dissemination of its principles through the sports media ecosystem, many other baseball teams—both professional and amateur—have recalibrated the very means by which they attempt to produce runs and winning outcomes.

One of the most interesting features of the sabermetric revolution in baseball and its dissemination through the flows of the media ecosystem is that much of the forensic analysis takes place outside the confines of the professional franchise. On web sites such as *Baseball Prospectus*, *Baseball Factory*, and *Sons of Sam Horn*, passionate amateurs comb through statistical archives to create and debate the metrics that may best measure the most rational means to manufacture runs and other components necessary to win baseball games. With this decentralization of actors engaged in sabermetric analysis it becomes very difficult for a professional franchise to create and maintain proprietary knowledge. Prominent sabermetric theorists, many of whom have made their reputations on the aforementioned blogs or on baseball discussion boards and listservs, are now consultants in MLB front offices, while every major mainstream sports media outlet now includes some form of sabermetric analysis in its product mix. As sabermetric ideas proliferate through the media ecosystem, the rest of the baseball world appears bent on closing the window of opportunity that Beane and the Oakland Athletics had opened for themselves.

TRANSVERSAL MOVEMENT: PROZONE AND SOCCER

There is a certain irony that a system which flourished largely due to relational databases and internet web sites was revealed to a broader public primarily through the print medium. But when considered as an entire media ecosystem, we see that the post-global network is much greater than a number of computers linked together by fiber optic cable and software protocol. The history of econometric analysis in sports is a history of such a networked ecosystem: Bill James' self-published print newsletter, originally advertised in a national magazine, becomes the archive from which sabermetric theory is developed; personal computers, spreadsheet applications and electronic bulletin boards foster the emergence of a community interested in the analysis of baseball statistics; NCAA college athletic department web sites allow the Oakland Athletics front office staff to cull data to implement sabermetric insights across a continental geography in advance of the entry draft; personal handheld computers similarly allow the franchise to implement sabermetric methods at all levels of management in the team's development system. And *Moneyball*, which uses the book form to reveal the methods by which Oakland achieved its success, is in turn supported substantively by web-based forms of promotion and distribution.

Understanding the case from such a perspective, it is not surprising that the application of econometric analyses to sporting labour did not persist as the exclusive domain of baseball. Enterprising managers in other professional sports began to explore how they, too, might incorporate such analytical techniques into their franchise operations and locate new efficiencies for competitive advantage. Despite this desire to apply econometric principles to other sports, however, we must first recognize that every sport is unique in its playing spaces, desired outcomes, rule systems, movement patterns and other structural conditions of possibility. Hence it may or may not be as possible in these cognate professional sport contexts to simply perform a regression analysis on existing quantitative data, as was done with sabermetric analysis in baseball, and determine price inefficiencies in the open market for athletic labour talent.

To illustrate this point and the implications it has for econometric analysis we shall turn from baseball to soccer, the world's most popular sport in terms of participation and spectatorship. Soccer, unlike baseball, does not consist of a series of discrete events segmented into easily identifiable units. Rather, it is an open-ended sport in which there are very few constraints on the flow of bodies within the side- and end-line boundaries of play, save the desired goal of putting the ball into the opposing team's net. This difference rests primarily in the structures of the respective playing spaces: the baseball stadium and its playing surface is relatively striated, while the soccer pitch, by contrast, has few enclosures and partitions of consequence, existing rather as a relatively open sea of grassy green expanse. As such, for the twenty-two players on a soccer pitch at any given time there are few spatiotemporal constraints on motion. We might contrast this space with the striated and call it relatively smooth, a space within which one navigates primarily by the line or trajectory of passage, subordinating the point in the process (cf. Deleuze & Guattari, 1987). Since there are fewer points in the playing space of soccer there are consequently fewer opportunities for measurement and as a result fewer quantitative data points available for the archival record, a marked contrast to the plethora of data that is produced and captured on a scoresheet during every inning of a baseball game.

Absent a rich statistical archive in soccer, the simple translation of baseball's analysis techniques from the sabermetric vernacular of OPS, VORP or Win Shares to a corresponding vernacular in soccer is not possible. But this has not dampened the enthusiasm with which soccer managers have sought to rationalize their sport and improve the efficiency of athletic labour expenditures on their individual teams. We might echo Eduardo Galeano's lament about the technocratic tendencies creeping into the beautiful game: "from the old blackboard to the electronic screen: now great plays are planned by computer and taught by video" (1998, p. 12). And so the principles of sports econometrics move transversally through the media ecosystem from the analysis and statistical modeling of archives in baseball to the video-based discrete object tracking of the ProZone system in soccer.

The use of film and video as a training tool in high performance sport has existed for decades. But it is only recently that computer and video have merged to provide the beginnings of an econometric intelligence such as that found in baseball. ProZone products and services, currently in use by over 100 clubs, leagues and associations worldwide, function by surrounding the soccer playing space with a minimum of 8 cameras, each calibrated for distance, angle of inclination, and other variables relative to a benchmark point on the pitch (ProZone, 2009). Each camera is furthermore synchronized with the others such that when the video feed produced by it is streamed via closed local area network to a central server and processor, it may be temporally and spatially aligned with the other video feeds. The computer may then synthesize these multiple points of view into a 360-degree model of the soccer pitch and the athletic bodies moving therein, such that each player on the field may be identified and isolated as unique data-objects that may be able to be tracked discretely despite the absence of any motion-tracking apparatus being affixed or inscribed on the athletic body proper.

There are two distinct modes of perception at work with the ProZone system. First, digital signal conversion techniques process the video feed from each camera by contouring what it perceives to be a unique body, and then interpolate and compare this contoured positionality with data feeds from the other cameras. In short, the system uses a haptic sense of perception to apprehend the athletic bodies operating in smooth space, no longer a singular locus of the perspectival gaze, but rather a constructed omnidirectional gaze that sees in full 360-degree relief. However, this positionality is made possible by a virtual grid that is laid over the soccer pitch by the ProZone processing software, creating a series of coordinate points that facilitate the identification, ordering and tracking of objects within the computer. In other words, the ProZone system virtually introduces a striation to the relatively smooth space of the soccer playing field, and we have discussed already the essentially optic nature of disciplinary space and its remote gaze.

As mentioned earlier, smooth and striated space are ideal types that exist everyday in admixture. With this continuous threshold of exchange between the two it becomes easier for the State to "utilize smooth spaces as a means of communication in the service of striated space." (Deleuze and Guattari, 1987, p. 385) The virtual grid that is overlaid upon the soccer pitch in the memory of the ProZone computer system transforms the space such that it becomes metric, insofar as it is able to synthesize and process multiple video feeds for the purposes of measurement. In this way the movements of the soccer player may be abstracted and integrated into the broader networks of the media ecosystem as objects of information. The athletes operating in smooth space who have been apprehended by haptic means now enter a striated space against which their performance may be measured for gains in economic efficiency.

The data produced by the ProZone system is available in real-time to the soccer clubs that subscribe to the service, and should be considered proprietary data locked into the conduits of a local area network for financial gain by sporting capital. But recently some of this data has been allowed to permeate into the broader realms of the media ecosystem: the Guardian newspaper in the United Kingdom has partnered with ProZone to make a limited subset of game data available to its web site audience on a time-delayed basis after games. Users may create their own "interactive chalkboards" analyzing and comparing players and teams in the English Premier League along a number of variables, such as pass efficiency and shots on target. These may be saved to the Guardian's online chalkboard community, embedded in personal blog posts or shared on social networking services. The concepts of econometric analysis in sport are thus spread beyond the bounded parameters of the local area network to the broader spectacles of post-global network consumption.

VISION AND TOUCH: THE FUTURE OF ECONOMETRIC ANALYSIS?

So we have two primary forms of data that have been leveraged to econometric ends in professional sport: the numerical event-based sort as witnessed with baseball and the rise of sabermetric analysis, and the image-based positional sort used in soccer and the ProZone application. But what if we were to merge these two types of data, combining the richness of event data with the visual elements of the image? Put differently, when the video feed becomes digital and thus exists as a complex bundle of binary code and linguistic meta-data, we can no longer speak of the image and statistical archive as different in kind. They are always already intertwined as forms of code distinct only in a qualitative sense, integrated in what Crandall (2003) refers to as a "body-machine-image complex" that serves to *track* bodies rather than simply observe them. As such, we should not be surprised to witness an intensification of the ways in which this complex of image and data is synthesized and incorporated into the practice of sports econometrics.

We witness the beginnings of such a synthesis currently taking place in the realm of professional basketball. Basketball is a far more open-ended sport than baseball, though less so than soccer. Similarly, its playing space is less striated than baseball, though far more so than soccer. As a result, basketball has more spatial opportunities for the capture of event data, while the openness, speed and physicality of the game increasingly demand the benefits of image-based analysis.

A *New York Times* story written, not coincidentally, by Michael Lewis (2009), details how Houston Rockets general manager Daryl Morey is attempting to bring econometric analysis to the professional ranks of the National Basketball Association (NBA), not unlike with Billy Beane in Major League Baseball. The story reads like *Moneyball* redux: a player with few discernible talents to the eyes of a scout or fan, nor possessing gaudy traditional statistics, in fact has quite a high value when econometric analysis uncovers latent drivers of efficiency from the folds of the sport. The difference in this NBA version of the *Moneyball* tale is that statistics alone only tell part of the story. Rather, it is the merger of statistical event data with video image data that enables Rockets management to index every play from thousands of NBA basketball games and extract meaningful intelligence.

The technology is provided by Synergy Sports Technology, a firm that provides a real-time video-indexing statistical engine and online retrieval platform for its NBA clients. It is in the process of patenting an "ontology-based tagging" system that allows professional basketball organizations to query databases of video and mine statistics in such a way as to better understand the relationships between players on the floor in specific game situations. Synergy Sports also offers a subset of this data to fans as part of its "Digital DNA" feature, which profiles "every athlete and sports team based on every move the players make in every game they play." (Synergy Sports Technology, 2009) With this sort of technology, Stross (2009) suggests, "professional basketball provides a working example of the unblinking eye that someday may hover above all workplaces."

While this is not yet quite the same thing as merging streams of event data with the multi-camera three-dimensional visualization of a system like ProZone, perhaps we are not far from such a scenario. Returning to baseball, one of the shortcomings of sabermetric analysis has been its inability to develop solid metrics for fielding, as compared to hitting and pitching. As mentioned earlier, there is a variability and fluidity in how the defensive players move on the field to record an out. The event of recording an out is significant, but does not take into account the spatial range with which one moved to catch a hit ball, since the patches of field between the bases would be considered relatively smooth spaces. Sportvision, a company that provides virtual graphic enhancements for televised sports broadcasts, has partnered with Major League Baseball Advanced Media, the league's Internet subsidiary, to create a system for baseball much like the one developed by ProZone (Schwarz, 2009). This system surrounds a baseball stadium with four high-resolution cameras that capture the action on the field and communicate with specialized software in a control room, which plots offensive players, defensive players and the ball in three dimensions, yielding about two million meaningful location points per game. This allows for some of the shortcomings of pure statistical analysis in baseball to be overcome through a combination of number and image.

It appears that the trend will only continue. STATS LLC., a company jointly owned by the Associated Press and News Corporation, and which bills itself as "the world's leading sports information, content and statistical analysis company," recently acquired SportsVU, a 3D visualization company offering products similar to ProZone (*Sport Business International*, 2008). It seems reasonable to assume that STATS will attempt to maximize this investment across various sports markets in conjunction with other products in its portfolio. More recently, STATS also acquired PA SportsTicker (the company that was established in 1909 as Western Union's original Baseball Ticker), a primary distributor of sports and information in the United States since the earliest days of telegraph communications. From a burgeoning network of telegraphy to the proliferating ecosystem of contemporary communications systems, an era seems to have passed in professional sport as a new one dawns.

The media ecosystem in which sports econometrics have flourished appears to have become at once archive and telesthesia, vector across time and vector across space, optic and haptic, located at different scales of geography, translated between diverse semiotic systems, and inscribed on or transported by various material substrates. Within this diversity, however, we find three consistently increasing tendencies: first, to abstract the embodied aspects of sporting creation and performance; second, to integrate the abstracted data and the performing bodies with advanced statistical models; and third, to relentlessly drive for efficiency in maximizing productive output while minimizing resource expenditure.

> The archiving, printing, writing, prosthesis, or hypomnesic technique in general is not only the place for stocking and conserving an archivable content *of the past*, which would exist in any case, such as, without the archive, one still believes it was or will have been. No, the technical structure of the *archiving* archive also determines the structure of the *archivable* content even in its very coming into existence and in its relationship to the future. The archivization produces as much as it records the event (Derrida, 1998, p.16, emphasis in original).

In essence, the local network of the stadium and the broader networks of sporting spectacle have made possible a thorough rationalization of the professional sport industry in assessing the relative value of individual labourers and towards the general intensification of productivity. But the modality of this rationalization is not that of the remote, perspectival panopticism in effect during the era of industrial modernization analyzed by Foucault (1977). Rather, material object, metadata, archive and network vector are integrated to form a relational assemblage at once visual, tactile and proximate.

Given that professional athletes earn such lucrative salaries and are continually in the media spotlight, it becomes easy for fans to continue considering them as "objects of information, never subjects in communication." As a result we become blind to the econometric techniques used in intensifying sporting labour that may someday soon be used to render all labouring classes more efficient. We ought to take heed of Derrida's cautionary note as the technical structures of media and communication continue to insinuate themselves into the sport production process: with the ongoing emergence and development of sport in the wires, the distinction between work and play is continually at stake for athletic labour, team ownership, fans, media networks, and the young athletes of today who will be the professionals of tomorrow.

References

Bale, J. (1994). *Landscapes of modern sport*. Leicester: Leicester University Press.
Baudrillard, J. (1994). *Simulations*. New York: Semiotext(e).
Castells, M. (2001). *The internet galaxy: Reflections on the internet, business, and society*. New York: Oxford University Press.
Castells, M. (2000). *The rise of the network society*. Vol. 1 of *The information age: Economy, society,*

and culture, 2nd ed. Malden, MA: Blackwell Publishing.

Crandall, J. (2003). *Drive*. Berlin: Hatje Cantz Verlag.

DeLanda, M. (1998). Meshworks, hierarchies and interfaces. In J. Beckman (Ed.), *The virtual dimension: Architecture, representation, and crash culture*. New York: Princeton Architectural Press.

Deleuze, G. & Guattari, F. (1987). *A thousand plateaus: Capitalism and schizophrenia*. (B. Massumi, Trans.). Minneapolis: University of Minnesota Press.

Derrida, J. (1998). *Archive fever: A Freudian impression* (E. Prenowitz, Trans.). Chicago: University of Chicago Press.

Foucault, M. (1977). *Discipline and punish: The birth of the prison* (A. Sheridan, Trans.). New York: Penguin.

Galeano, E. (1998). *Soccer in sun and shadow* (M. Fried, Trans.). London: Verso.

Guttman, A. (1978) *From ritual to record: The nature of modern sports*. New York: Columbia University Press.

James, B. (1988). *The Bill James historical baseball abstract*. New York: Villard Books.

Levinson, P. (1997). *The soft edge: A natural history and future of the information revolution*. London: Routledge.

Lewis, M. (2009, February 13). The no-stats all-star. *New York Times*. Retrieved from http://www.nytimes.com

Lewis, M. (2003). *Moneyball: The art of winning an unfair game*. New York: W.W. Norton and Company.

Lowes, M. (1999). *Inside the sports pages: Work routines, professional ideologies, and the manufacture of sports news*. Toronto: University of Toronto Press.

Marazzi, C. (2009). *Capital and language: From the new economy to the war economy*. New York: Semiotext(e).

Markula, P. & Pringle, R. (2006). *Foucault, sport and exercise: Power, knowledge and transforming the self*. London: Routledge.

McLuhan, M. (1964). *Understanding media: The extensions of man*. New York: Harper.

National Baseball Hall of Fame. (n.d.). *Dressed to the nines: A history of the baseball uniform*. Retrieved April 10, 2006, from http://www.baseballhalloffame.org

Parikka, J. (2007). *Digital contagions: A media archaeology of computer viruses*. New York: Peter Lang.

Postman, N. (1992). *Technopoly: The surrender of culture to technology*. New York: Knopf.

ProZone. (2009). Retrieved from http://www.prozonesports.com

Puerzer, R. (2002). From scientific baseball to sabermetrics: Professional baseball as a reflection of engineering and management in society. *NINE: A Journal of Baseball History and Culture*, *11* (1), 34–48.

Quirk, J. & Fort, R. (1997). *Pay dirt*. Princeton, NJ: Princeton University Press.

Raney, A. & Bryant, J. (2006). *Handbook of sports and media*. Mahwah, NJ: Lawrence Erlbaum Associates.

Schwarz, A. (2009, July 9). Digital eyes will chart baseball's unseen skills. *The New York Times*. Retrieved from http://www.nytimes.com

Shaviro, S. (2003). *Connected, or what it means to live in the network society*. Minneapolis: University of Minnesota Press.

Shogan, D. (1999). *The making of high-performance athletes: Discipline, diversity, and ethics*. Toronto: University of Toronto Press.

Sport Business International. (2008, February 12). STATS acquire SportVU. Retrieved from http://www.sportbusiness.com

Stang, M. and Harkness, L. (1996). *Baseball by the numbers*. Lanham, MD: Scarecrow Press.

Stross, R. (2009, April 29). Technology to dissect every dunk and drive. *The New York Times*. Retrieved from http://www.nytimes.com

Synergy Sports Technology. (2009). Retrieved from http://www.synergysportstech.com

Tygiel, J. (2000). *Past time: Baseball as history*. New York: Oxford University Press.

Wark, M. (2004). *A hacker manifesto*. Cambridge, MA: Harvard University Press.

Wenner, L. (1989). *Media, sports and society*. Newbury Park, CA: Sage.

CHAPTER NINE

Television as Everyday Network of Government

JAMES HAY

TELEVISION/NETWORKS AND STUDIES OF GOVERNMENTALITY

Armand Mattelart has traced the Modern "discovery" of the network, and its relation to the "invention of communication," to multiple and intersecting sciences devoted to the health of circulatory systems which developed between the seventeenth and nineteenth centuries. He notes that the pre-Modern use of the term "network," which signified lace-making, was rearticulated in the seventeenth-century by Italian naturalist and physician, Marcello Malpighi in reference to the flow of blood through arteries. He points out that medical science's rationalization of the human body as a circulatory system, and its conception of healthy bodies as free-flowing circulatory systems intersected with contemporaneous conceptions of "economy" by French Physiocrats, the increased reliance on systems of weights and measures to chart healthy flows, and the mapping and engineering of roads and river routes. "Communication" as healthy, free-flowing network became an object of inquiry, knowledge, and invention along these various "paths of reason." I begin with this reference in part to underscore that the history of the communication network is old (albeit Modern) and can not be understood merely as a history of communication media—unless of course one is willing (as is Mattelart) to recognize the historical genealogy of "communication." This is a lesson that becomes more prescient with every exclamation that we are living with "new media" in a "network society."

Throughout much of the twentieth century, "network" referred to a particular formation and arrangement of communication described by terms such as "mass communication" and "broadcasting." Broadcasting was organized as national territories/zones of communication through radio (after the mid-1920s) and television (after the late 1940s). Broadcasting was financed or subsidized differently in different parts of the world (as "public" or "private" corporations), and its mission as popular cultural form became central to the production of national cultures. In the United States, broadcasting occurred through the linking of local facilities into national networks. The National Broadcast Company (NBC), a subsidiary of the Radio Corporation of the America (RCA) which manufactured radio and phonograph products, became the first national broadcast network in 1926, followed by the Columbia Broadcast System (CBS) two years later and by the American Broadcast Company (ABC) in 1943. By the early 1950s, these three broadcast networks also operated as the three distributors of television programming in the U.S. The creation of the Corporation for Public Broadcasting in the late 1960s was rationalized as an alternative to the commercially driven objectives of the other three TV networks, though it too was structured as a national network of local "public broadcast" stations.[1] From the 1920s through at least the 1970s, national broadcast networks operated (albeit differently across the world) as arrangements for shaping and managing political, economic, cultural, technological, gendered, racial, and social citizenship. While these networks—this conception and organization of network—were not the only means of shaping and managing citizenship, the forms and models of citizenship that became dominant in this period relied profoundly on the network as a means of linking everyday life what counted as civics, *polis*, and "government."

Various expressions have been adopted recently to describe the passing of this twentieth-century meaning and arrangement of "network." The term "post-network era" figures prominently in recent considerations about television's transformations. Amanda Lotz, for instance, embraces the term (albeit cautiously) as a useful way to consider the demise of "the network era" of TV, "when 'television' meant the networks NBC, CBS, and ABC." (Lotz, p. 11) Lotz rightly charts the emergence of the "post-network era" through the proliferation of cable and satellite distribution of TV after the mid-1970s (a passage from broadcasting to what some have referred to as "narrowcasting"), and she qualifies the epochalist intonation of "post"-ing a "network era," stating that her use of the term is not intended "to suggest the death or complete irrelevance of what we have known as television networks and channels." (p. 254) However, her account of a "post-network era," focuses on television as a relatively *discrete* industry, cultural form, technology, network and history, concludes that television is still (somehow) a discrete network: "Decades later, we no longer need a separate medium to frame our understanding of television because its own historical features and distinctions now serve that function." (p. 256) Should we accept those "historical features and distinctions" as if they simply were self-generated? The final paragraph of her book quotes Nicholas Negroponte's statement (from 1995) that "the future of television is to stop thinking of television as television" (p. 256), but it is difficult to imagine from her account how to analyze TV by first de-centering TV as a self-perpetuating entity. Although this is a fairly insignificant point to dwell on in one respect, it bespeaks a disposition about what determines the history/transformations of a single *medium* such as TV and (as discussed below) it still figures into current assumptions about "media power."

Lotz rightly notes that her use of the term "post-network era" is in some ways contradicted by the recent attractiveness of the expression "the network society" adopted by followers of Manuel Castells' eponymous book (Castells, 1996). Castells' conception of the "network society" is concerned more with how networks operate as distributive apparatuses of "information" than with the legacy of broadcasting and its role in producing and distributing entertainment (Lotz's focus). He therefore is less focused on the specificity of networks supporting a specific medium than is Lotz. Whereas Lotz concentrates on the new strategies of production, distribution, and audience measurement in an old broadcast medium, Castells is preoccupied with the new forms of productivity that rely on information technologies and networks and whose emergent footprint—a "space of flows"—favors certain centers of productivity while marginalizing other areas. However, as Lotz acknowledges, her use of the term "post-network era" is not incommensurate with Castells' term "network society" because they both describe a historical rupture hastened by a widespread reliance on "digital media," the Internet, and World Wide Web, whose pervasiveness in contemporary life is purportedly so thorough as to warrant the expression "network society." For both Lotz and Castells, Nicholas Negroponte's account of the "digital revolution" lays historiographic and theoretical groundwork for how media matter in the present because Negroponte described (and lauded) the "digital revolution" not only as breaking up broadcasting companies' and video cassette retail companies' control of material distributed through their networks, but also as forging a "post-information age" multi-media environment— a wedding of broadcasting and interactivity.

One solution to explaining the ramifications of the "digital revolution" and its role in transforming broadcast networks such as television's has occurred through the expression "convergence culture." Mark Deuze applies Castells' rationale about the network society (rather uncritically) to map the new centers and geography of various "creative industries" while emphasizing (more than Castells) how media play an increasingly central role in the everyday life (the "media life") and productivity (work and play) of consumers. Echoing a rationale advanced by Henry Jenkins, Deuze proposes that "the new human condition [of the network society], when seen through the lens of those in the forefront of changes in the way work and life are implicated in our increasingly participatory media culture, is convergence. . . . Media convergence [has] a cultural logic of its own, blurring the lines between production and consumption, between making media and using media, and between active and passive spectatorship of mediated culture." (Deuze, p. 74) That said, Deuze (like Castells and Lotz) attributes the current media convergence primarily to an economistic logic specifically driving the production, distribution, and commercial synergy of multiple media. He argues that this logic is worth mapping because these media networks and technologies have become central to contemporary life (our "media life"), though he examines the networks of interactive *consumers* and "media life" mostly in terms of their value within the new geography of creative industries.

Jenkins' argument about a new "convergence culture" most energetically addresses the tension (or arrangement) between media industries' strategies for encouraging consumer participation and consumers' (particularly the most invested gamers' and fans') active involvement and *interactivity* with the production and life of media commodities. Noting that "convergence" is "both a top-down corporate-driven process and a bottom-up consumer-driven process," he states that "media companies are learning how to accelerate the flow of media content across delivery channels to expand their revenue opportunities, broaden markets, and reinforce viewer commitments [while] consumers are learning how to use these different media technologies to bring the flow of media more fully under their control and to interact with other consumers." (p. 18) Although Jenkins implies that the *culture* of convergence is produced along both these paths, he occasionally poses the "grassroots" and "bottom-up" attributes of "convergence culture" as a counter-pressure to the commercialist strategies of "big media" industries. This latter impulse in his account of contemporary convergence culture is particularly evident when he calls-out veins of critical studies about media (e.g., the work of Mark Crispin Miller, Noam Chomsky, and Robert McChesney) that have not adequately recognized how the non-professional, consumer-driven practices of "collective intelligence" and a "participatory culture" complicate their views of media power, which he sees as too oriented toward older models of media production and distribution. Though his *Convergence Culture* begins by charting the changing uptake of Negroponte's optimism about a digital revolution, he concludes his book insisting (as did Negroponte at the end of *Being Digital* ten years earlier) that "increasing participation in popular culture is a good thing" and something too often ignored by "critical pessimists." (p. 248)

To the extent that Jenkins ties the virtue of "participation in popular culture" to the "democratic" possibilities of a new "convergence culture" (i.e., to "the politics of participation"), media convergence is either cast as a field of political intervention overlooked by "critical pessimists" or inherently more democratic than was the broadcast era—a time purportedly *before* "convergence culture." Whereas Lotz's account of the "post-network era" is conducted under the slogan, "The Television will be Revolutionized," Jenkins' account of "convergence culture" reassesses the slogan, "The Revolution will not be Televised," noting that the most significant forms of political mobilization since the 2004 presidential campaign have occurred through non-professional (fan and avid consumer)-driven techniques such as Photoshopped political ads and through "grassroots" networks of activism and intervention. In this sense, TV's convergence with "new media" (its place in emergent media networks of political activism, or maybe a "network society"), and TV's engagement by consumers who are savvy about using these media/networks for personal and collective self-representation, have demonstrated what makes the current media convergence more democratic than in the past.

Although Jenkins is right to point out that today democracy is represented and practiced *through* the media technologies and networks that he associates with a "convergence culture," his argument tends to veer toward a universalist conception of liberal democracy and its flowering or return in a "participatory," "grassroots," non-professionalized media culture. It is one thing to point out (as does Jenkins) that "consumer communities" and media fandom involved in "grassroots" organizing helped galvanize political candidates such as Howard Dean in the 2004 presidential campaign, and another to interpret that as an indication that "community networking" simply enhances citizenship and makes for a purer form of democracy than in the past—or presumably in other parts of the world lacking these practices. Although Jenkins gestures (around the edges) to the ways that television and its linkages to web-based consumer-citizen communities can be described as what John Hartley has termed (in reference to current TV series such as *American Idol*) "democratainment" as a commercial initiative, he does not push hard enough at the contradictions surrounding the shaping and valuing of energized consumers and their "communities" (for instance, their complex role in reproducing an economy of interactivity and mass customization, as Mark Andrejevic has argued).

Without ignoring the contradictions of the agency of citizen-consumers within the economy of media "interactivity" and "participation," I want to propose an alternative to economistic accounts of media networks (e.g., Lotz, Castells, Deuze) and Jenkins' valorization of a "convergence culture" and its consumers as ushering (driving) a more democratic form of citizenship. Explaining this alternative involves moving beyond the starkly binaristic logic that pervades these accounts: ownership vs. consumption, political economy vs. (fan) culture, passivity vs. inter-/activity, control vs. resistance, government vs. community mobilization, and "critical pessimism" vs. "critical utopianism" (Jenkins' terms) as the only viable positions for explaining "media power" in the "network age" (or even a "post-network age"). While Jenkins is right that it is important to understand questions of media power, political agency, and citizenship (not just, in his terms, "the public's role in the political process") by considering how technologies and networks are produced and reproduced through "the everyday life experiences of citizens," (p. 208) one still needs to consider certain contradictions surrounding the relation of media/citizenship to the *technologies and networks of government* through which citizenship is shaped and directed in their everyday lives.

Moving in this latter direction involves, in part, rethinking Michel Foucault's writing about subjects, power, governmentality, biopolitics, technologies of the self, and neo-/liberalism. There are several implications of Foucault's work around this cluster of topics that the following sections of this chapter consider. First, Foucault repeatedly emphasizes that freedom is not the opposite of social control. His account of liberal government demonstrated how the "birth" of institutions oriented toward the sovereignty of the individual and toward personal liberties were accompanied in the late-18th and 19th centuries by the proliferation and dispersal of mechanism for disciplining individuals and populations and for guiding and shaping the proper, healthy exercise of freedoms. For Foucault, liberal government was initially shaped by the tension between these two developments, and it subsequently reinvents itself in response to potential resistances and particular problems in governing this way.

Second, in considering how programs of empowerment, healing, and social care have operated as technologies of disciplinarity, and how the achievement of individualism and personal freedoms occurs through the *proliferation* and dispersion of the technologies of self-discipline, social welfare and security, Foucault emphasized the importance of recognizing the "micro-physics" and "micrologics" of power—the little, everyday operations of social control. In his writing about "governmentality," he famously underscored that an analysis of liberal "government" (political modernity) did not begin and end by examining the operations of the liberal State but instead by mapping the multiplicity of the techniques and technologies of control which are administered by community, family, individuals, etc., on themselves. His view that government is practiced/enacted through everyday caring for and administering to community, family, and oneself implies that citizenship is not simply one's relation to the State but also the multi-form and daily ways that *life* is managed.

An analytic of government thus maps the many points of application and many varieties of technologies through which control and freedom are exercised, but also the ways that government continually anticipates and develops technologies for overcoming resistances. It maps how power is exercised through networks, and dispositions within these networks. Government depends on historical and geographic "arrangements" of these networks and dispositions. The longevity of liberalism has been predicated on the anticipation of and experiments in circumventing potential breakdowns in arrangement of government, and on finding the proper and most healthy arrangements of governing through freedoms (what Nikolas Rose, after Foucault, has term "the powers of freedom"). Whereas Mattelart charts the historical coalescence of reasonings about healthiness of unblocked, freely flowing circulatory systems and economies, Foucault underscores that these reasonings and networks became instrumental to the dispersion of the mechanisms and points of "liberal government."

What questions can be addressed by considering communication media through an analytic of the networks, technologies, and arrangements of government? Foucault's relative lack of attention to communication media and networks actually is useful in thinking about media practices because, in suggesting that power lacks a center (i.e., that it is dispersed), he proposes an analysis that must figure out how, when, and where media matter within networks and arrangements of government. That he posed his account of power and governmentality as an alternative to a Marxist political economy and to French Structuralism (the theoretical framework for ideological and deconstructionist criticism), also suggests an analytic that raises questions not typically asked in critical studies of media which have long preferred these theoretical frameworks for explaining media power. To the extent that an analytic of the everyday networks of government emphasizes the diffusion and popularity of certain sciences, protocols, and technologies of government and citizenship, it sees communication media as activated through governmental and managerial rationalities that are not limited to the communicative functions of media (and thus the ways that communication becomes, particularly for Communication Studies, the basis for understanding every question power). An analytic of everyday government thus cuts two ways: as an analysis of media industries as well as consumer-citizens who are reliant on *techniques of management*, and an analysis of how these privatized and personalized techniques of management operate in conjunction (or not) with policies, programs, and regulations of the State. Historically and geographically these techniques, and the networks amongst these agents, has varied—with greater and lesser degrees of a role by private (as distinct from State) agents of management. Historically broadcasting, narrowcasting, and media interactivity have been integral to liberal government in the U.S. through these changing arrangements of management and government; networks of everyday government act on and through these regimes of communication media, though not primarily or exclusively through them. In that the recent stage of liberal government (a so-called "neo-liberalism"), which has encouraged greater reliance on privatizing public services, has intersected with a "post-network era," a "network society," and a "convergence culture," it is important to understand how media such as television operate within this governmental arrangement and rationality as a perspective as much about current liberal government as about the current convergence of media.

In proceeding this way, my chapter taps and rethinks the important literature about television's relation to everyday life–studies that emerged since the 1980s, amidst the decline of the broadcast model and its relation to a stage of liberal government. I examine how television (in its relation to other media) is instrumentalized as technology and network of everyday government and through rationalities and programs oriented to governing everyday life, producing and authorizing particular forms of citizenship. While television or other Modern communication media have never been discrete networks of government, their operation within governmental rationalities has imbued them with relatively specific objectives in daily life. Programs and rationalities of government also have objectified them as the source of particular problematic behaviors in need of correction (e.g., TV as having spawned a nation of "couch potatoes"). It may be useful to conceptualize television (as an industry, technology, or cultural form) that was separate from cinema and radio during an age of broadcasting, or to discuss the convergence of contemporary media; however, it is just as important to develop an analysis of media/power that situates the practice and virtue of separation and convergence, as well as the spirit of invention and modernization associated with "new media," within rationalities and programs of government and citizenship. Citizenship and government are enacted daily, *partly* through media technologies and networks, and certainly not only as the result of media consumers' information-gathering or monitoring about public/political affairs.

REINVENTING TELEVISION AS A NETWORK OF GOVERNMENT: DOT.GOV THROUGH DOT.ORG AND DOT.COM

What is liberal citizenship through the current arrangements of managing life everyday, and how do communication media matter as technologies of citizenship and government? I want to address these two questions by focusing on television not only because, as demonstrated by nearly all of the authors whose views I have cited in the first section (with the exception of Castells), the broadcasting model of television has been repurposed within an environment of convergence, but also because the current discourses about "media convergence" almost never pose questions such as these. Without setting aside television's technological reinvention, its cross-media commercial alignments and strategies, and its encouragement/requirement of greater consumer interactivity, this section (and the one that follows it) take up three considerations for thinking about how television matters within the current rationality about liberal government and citizenship.

The first consideration has to do with television as a *network of government*. As Laurie Ouellette and I have noted, television's reinvention within the current media convergence has occurred through a governmental discourse and policy initiatives that valorized "the reinvention of government" as a basis for a broad rethinking and remodeling of the "welfare state." "Reinventing government" is not simply our own figure of speech; it is an expression that Bill Clinton absorbed in 1992 from David Osborne and Ted Gabler's book (from that same year), *Reinventing Government: How the Entrepreneurial Spirit is Transforming the Public Sector*. The term was acted on—developed, under certain historical conditions—as a "discursive formation" and as integral to a new ("neo-liberal") rationality about government and citizenship. The impetus to "reinvent government" rested upon a paradoxical aspiration of "neo-liberalism" by economists and managerialists such as Peter Drucker who, in the wake of resurgent interest in the writings of economist Freidrich von Hayek, cited Nazi Germany and the Soviet Union as evidence of a malaise resulting from the "diminishment" of individual freedoms at the hands of government intervention. For Drucker and von Hayek, liberalism needed to be *renewed* by "returning" to the practices that preceded what they viewed as expanding bureaucracies and the swelling size of public welfare programs.

This vision of "the entrepreneurial spirit" transforming the public sector through "public-private partnerships" informed various "reforms" and policies, the reasoning about government's reinvention both deepened dramatically and took slightly new directions during the Bush administration. Its Faith-based and Community Initiatives, the USA Freedom Corps Volunteer Network, the Volunteers for Prosperity, the Citizen Corps, and Homeland Security all developed out of (and sometimes away from) Clinton's National Partnership for Reinventing Government, initiated in the first year of his presidency. During the Bush administration, federal contracts with industry doubled from just over two hundred million in 2001 to over four hundred and fifty million in early 2008, with outsourcing becoming a normal practice for State agencies from the Forest Service to the CIA.[2] Rearticulating the connotations of the idea of "public service," i.e., a service provided by the State but also a citizen's or company's responsibility to provide social assistance through not-for-profit programs, thus became a fundamental objective of this governmental rationality.

The virtue of the "public-private partnership" has a longer history in U.S broadcasting than in other countries whose system of broadcasting developed as State-subsidized public services. Certainly U.S. television's long history of commercially sponsored broadcasting operated as a fertile basis for rationalizing at the end of the twentieth century the virtue of privatized provision of social welfare. But from the early years of TV broadcasting in the U.S., television had played a role in shaping and rewarding proper forms of citizenship, as in *Mr. Citizen* (1955) which showcased ordinary people who came to the assistance of needy individuals and who were given a "Mister Citizenship Award" by a "prominent American." There are other examples throughout U.S. television history of series that designed specific broadcasts in support of a specific federal program. And, as Ouellette has explained, the creation of a Public Broadcasting Corporation in the late 1960s launched a television network that governmentalized the cultural instruction necessary for healthy citizenship (Ouellette, 2002). However, it is no small coincidence that recent television (particularly the Reality TV syndrome) and the forms of media programming and convergence which support it developed within the discourse and policy-initiatives about "reinventing government," and with considerable energy during the Bush era. A prominent vein of the post-broadcast era involved the proliferation of instructional programming in networks such as Discovery Channel, The Learning Channel, Home & Garden Network, the Weather Channel, and The Food Channel, which are devoted entirely to specialized topics, and in numerous other cable networks which are not primarily devoted to instructional formats. More than in the past, this vein of television's linkage to web resources has made it a vital part of the networks' sustaining the public-private partnerships and their new arrangement of "public service" through which liberal government has staged its "reinvention."

One of the most durable Reality TV series since early in the Bush administration has been *Extreme Makeover: Home Edition* (*EM:HE*), which premiered in the spring of 2003.[3] The series selects individuals who are judged "worthy" of recognition typically because they are paragons of care-giving for family, neighborhood, or community, and because their form of private, individualized social service could be enhanced had they the resources which the series provides them. Usually the series' provision of assistance involves the renovation of their house, often because the house doubles as a setting for privately administered care-giving—the household as a zone of welfare between business and charity, as when *EM:HE* refurbishes the property of the Hill family who has used their farm as a boxing training facility for underprivileged youth. The series' material rewards to its subjects always are represented as philanthropic—corporate philanthropy rewarding/assisting unrecognized and deserving providers of welfare.

Some episodes reward individuals who already are working with government sectors such as the military that are being transformed through increased reliance on "outsourcing." The series' second episode began this trend by staging a "home makeover" for a family comprised of a father whose National Guard unit had been sent to Iraq. At a time when furloughs for enlisted personnel were non-existent and National Guard units were being activated because of a shortage of regular army units, the series' producers struck an agreement with the Defense Department to permit the soldier-father to return in order to participate in improving a house into which his family had just moved before his activation to Iraq. In an episode involving the Cooper family, the series adds their own corporate "honors" on a disabled veteran of an Iraq campaign who has been "working with Congress" to improve veterans benefits.

As Lotz notes, television's "post-network era" has involved a realignment of local stations to the national networks. Whereas the four national networks formed licensing agreements with local affiliates, local TV stations now have begun to venture into new strategies that make them less dependent on the national networks, as cable and satellite networks rely on cable providers rather than local networks. However, series such as *EM:HE* which are distributed through one of the old national networks also assume a relation to localities that is congruent with the federalist orientation of the "neoliberal" governmental rationality. Often rationalized as respecting "states' rights" and as "freeing" localities from national "bureaucracy," federalism valorizes the importance of decentralizing governmental administration, not only cultivating corporate administers of public services but also off-loading administration onto states and municipalities. The Bush administration's response to the victims of Hurricane Katrina in 2005, for instance, cast the federal government as "supporter" of state and local responders and non-profits such as the Red Cross, eventually outsourcing tasks to private contractors. Through this governmental arrangement, *EM:HE* produced an episode just before Christmas 2005 in which Laura Bush (having declared publicly that *EM:HE* was her favorite TV show) accompanied the usual cast of the series' renovation technicians to Mississippi to rebuild a house which had been damaged by the hurricane and which required special facilities for the family's Down Syndrome son. The series emphasizes its role in mediating the organization of and compensation for the army of local, "grass-roots" samaritan-providers, as its website represents linkages to local non-profits or local chapters of national ones.

Mostly through its website, ABC touts *EM:HE* as the ridgepole of its Better Community initiative, providing an archive of links to descriptions of past episodes as well as the non-profits and public service providers that either were represented in the episode or whose relevance to a type of relief the episode demonstrated. For instance, the Better Community links for the episode about the Gaudet family whose house was damaged by the hurricane include the

United Easter Seals, United Cerebral Palsey, the National Association for Down Syndrome, and the Down Syndrome Society of Mobile County (where the episode was set). Links for the Cooper family episode include Disabled American Veterans, the Warrior Intern Network, Helping a Hero, and Homes for Our Troops. Although Lotz and Jenkins would be right to note that, characteristic of the cross-media merchandising and embedded promotional strategies of contemporary TV, *EM:HE* rewards its exemplary citizens with products provided by the advertising sponsors of the episode (such as Sears/Kenmore products), even these product references are stitched into the series' demonstration of the philanthropy that operates within the governmental rationality sanctioning public-private partnerships. The Better Community website's list of participating and "spotlighted" organizations (described in the website as its "partners") includes a link at the bottom of the list for its parent-company Disney's "outreach" initiatives.

While the website purports that the series is one engine for "building a better community, one family, one house, one donation at a time," the website represents a good and healthy community as a network of "partnership" between non-profits and corporate-sponsored welfare. Materially, television is instrumentalized within that network. More than the TV series, the website collectivizes, mediates, and valorizes a network of privatized public service, though the website's capacity to enact network as private partnership (a "better community") depends on this vein of Reality TV, which provides a technical *demonstration* of the shaping of productive citizenship within the Better Community network of web-based linkages. *EM:HE* belongs to Reality TV genres (on ABC and other TV networks) that have privatized the provision of various forms of welfare and catalyzed a civic responsiveness (Better Communities and their good citizens) through these network-partnerships, though the generic conventions of these TV series are not simply representational; they include a set of technical conventions such as web-links and interactive mechanisms—a convention found in the former ABC series *Oprah's Big Give* and its website's links to her Angel Network of social helping, or Fox's *American Idol Gives Back* and its website's links for donations and to its network of corporate partners.

Typically, the link between State and privatized televisual administration of public services is not formalized; indeed its formalization would contradict the governmental reasoning and arrangement that accept outsourcing and "partnership" as a natural extension of the historically limited role of the State in U.S. broadcasting and "free TV." The NASA channel thus operates within NASA's longstanding hybridity/ambiguity as a State program heavily dependent on contracted labor, services, and research. The Weather Channel's claim to be "the nation's premier provider of weather information" (http://www.weather .com/aboutus/?from=footer) relies on and sometimes invokes the authority of the National Oceanic and Atmospheric Administration and the National Weather Service (both overseen by the U.S. Commerce Department), but only during crises (extreme weather and emergencies) does the Weather Channel foreground and formalize its relation to the National Weather Service, or do National Weather Service broadcasts pre-empt or visually become part of (through flashing warnings and banners) Weather Channel broadcasting. The "non-profit" C-Span (subsidized and overseen by a consortium of cable companies) operates through an arrangement to broadcast, and more recently distribute on-line, congressional proceedings. And as federal courts have slowly begun to allow the presence of video cameras in courtrooms, the TruTV Channel (formerly Court TV) has produced and broadcast programs about court cases, frequently using extended video footage from courtrooms and police case work.

Though there are numerous examples of these private networks providing a public service or welfare, their status as public servicers can not be attributed only to their ubiquity in replacing or reinventing services once administered by the State, in however limited a capacity that once occurred. The rationality of a public-private partnership emphasizes the "supportive" role that the State is supposed to play–not being the primary provider of welfare but encouraging, rewarding, and (in that way) stitching itself into and acting on private networks of social service. In addition to Laura Bush's appearance on *EM:HE*, the Bushes appeared in a supportive posture as part of the Idol Gives Back broadcasts.[4] In that the websites for State agencies during the Bush administration such as the Department of Homeland Security, the program for Cyber-security, and the Department of Health & Human Services provided links to private "partners" (corporations and non-profits), these networks linked State networks to partners also identified on the television websites. In that sense, the TV linkages can disavow their relation to the State (a longstanding disposition of "free television"), while reproducing a network of provision and government. There certainly are significant examples of the State's initiation of the public-private partnerships involving television and related media. The Bush administration's Homeland Security created "information networks" with localities and "private partners,"[5] and the Bush Economic Development Agency staged "telecasts" over "government networks" of video demonstrating the possible interfaces of public and commercial entities in networks of "economic development." More commonly, however, Reality TV productions such as the ones cited are the *creative* engines of public service, designing and performing the partnership, as the State inserts itself into, and acts through, these private (commercial or non-profit) networks of welfare and government.

While the Department of Homeland Security's "outreach" once relied upon news services to disseminate its infamously color-coded alerts, its web site during the Bush administration was its primary means of relaying citizens to federal, state, municipal, private, non-profit, and self-services. In January 2009 ABC's *Homeland Security, USA* (*HS,USA*) fashioned the U.S. as *televisual* "homeland" administered as a partnership between ABC and the Department of Homeland Security, and as a product and market that are managed by both the ABC and Homeland Security brands. ABC's website for its *HS,USA* claimed that the series is unprecedented because its producers were granted special access to various Homeland Security agencies and because it transports viewers to the "front lines" of these agencies' operations: "The epic landscape" of "America's [sic] borders . . . , a territory that includes airports, seaports, land borders, international mail centers, the open seas, mountains, deserts and even cyberspace." The map of the nation offered by *HS, USA* was, at least in the initial episodes, an interminable border region—a zone that deepened or reinvented Reality TV's games of surveillance (of being watched and watching oneself) and Reality TV's citizenship games (tests and trials of belonging or being kicked off the island). This map represented the U.S. as a gated community whose network of gateways everyday are under little assaults (on-going games and strategies of stealth) that require equally stealthy monitoring and regulation (sniffing out clues on the game board). Like other Reality TV series that have mapped the nation through *popular* contest (controversial series such as ABC's *Welcome to the Neighborhood*, more successful ones such as *The Amazing Race*, or national events such as *American Idol*, *HS, USA*'s competition at that hour), *HS, USA* deployed its own panel of experts who decided who can pass—who wins and who loses the citizenship game.

While *HS, USA* was a form of "publicity" (advertisement and public service information) or even, in stronger terms, propaganda for Homeland Security, these descriptions rest on complicated and changing relations between the residual forms of citizenship left over from television's past as a broadcast medium, and the self-directed forms of citizenship accompanying television's reinvention through the interactive economies of web-sites. They also rest on the emergent way that the technologies for *membership* in private "communities" (whether gated communities or on-line varieties such as those routinely provided on TV websites like ABC's and *HS, USA*'s) are becoming the technologies of citizenship in sovereign territories.

A TV franchise that most vividly and notoriously underscores the historical contradictions of television's production or mediation of the public-private partnership was the NBC and MSNBC series, *To Catch a Predator* (2004–2008). The series worked with a non-profit on-line watchdog foundation, "Perverted Justice," which monitors "predatorial" behaviors over the Internet, and with state and municipal law-enforcement agencies, in order to stage sting operations which lure men to the house of an underaged girl-agent in order to elicit an on-camera confession which will result in the man's arrest. The staging of these entrapments, confessions, and arrests involves Perverted Justice's reporting suspects to the show's producers, and the producers arranging with local law enforcers to hide at the house in order to secure the premises and subsequently make the arrest. The series' producers made arrangements with law enforcement agencies in numerous localities in the U.S. during the life of the series. In all cases, the local government became a partner in financing the staged event, having to pay police to participate in the sting operation. In some cases, the series' recorded material became evidence in subsequent prosecutions of the entrapped offenders—in some instances leading to convictions and in other instances leading to the case's dismissal because of the indicted individual's entrapment. Indeed the series represented the contradictions of a federalism which assumed the efficacy of public-private partnerships—a national TV production mobilizing and sometimes undoing local rules and their local adjudication.

The producers of NBC's *Dateline* followed *To Catch a Predator* with the series *The Wanted* (July 2009) which, according to NBC's press release for the series, assembled "an elite team with backgrounds in intelligence, unconventional warfare and investigative journalism" in their quest to assure that "justice is served" to suspects designated as terrorists by the Bush administration. The group's on-screen organizer professes to have been part of a private association tracking terrorists internationally, and the other members of the group are former U.S. Army and Navy special operations soldiers or employees of State agencies—all now working as private players (in and out of Reality TV). Their quest involves presenting evidence to (and badgering) foreign governments whose systems of justice are made to appear to have failed in apprehending suspects living under their jurisdiction. Although the series' *serving* of justice does not ostensibly coordinate with agencies of the U.S. government, the series is perfectly compatible with the trend toward outsourcing during the Bush administration's "war on terror" and in the current regime of global governmentalities. The series operates within a network of mediation between judicial and policing agencies and private agents whose precedent is established as much by prior TV franchises such as *To Catch a Predator* as by the governmental rationality sanctioning the outsourcing of international security enforcement through companies such as the Blackwater/Xe Corporation (paid millions of dollars by the Bush administration as a security force to aid the U.S. Army in apprehending "terrorists" in and "securing" occupied Iraq).

Thinking about Reality TV's networks of government through these examples highlights how the *realism* of the Reality TV productions has to do not simply with their documentarist aesthetic but with their technical demonstrations (through a network linking television and web resources) of public-private networks of government, and with their creative/generic design of the proper (and most lucrative) networks for mediating these partnerships. In that sense, Reality TV is not so much an example par excellence of a "postmodern aesthetic's" blurring of the difference between the Real and its representation as it is the production of the networks of government as public-private partnership. To the extent that these TV productions are "game-docs" or "docu-dramas," they confer an institutional history of the technical objectivity ascribed to the State agencies of liberal democracies (entities of rules and laws for making rationale and objective government administration) onto various private and non-profit agents of government. As demonstrations and exercises in government as public-private partnership, the productions also are experiments (tests and contests) for demonstrating that privatized networks are viable media for administering the services associated with public institutions. In this dynamic, the work of the private and the State agents objectify (take as their object and act through) the rules of one another's operations.

THE PERSONALIZATION & EVERYDAYNESS OF TV'S NETWORKS OF GOVERNMENT

While the prior section dwells on some of the ways that television has been reinvented to mediate dot.gov, dot.com, and dot.org networks (networks of private government and welfare), this section briefly addresses how television's reinvention and place in the current networks of government depend on and increasingly require individualized, customized technologies for managing/governing various aspects of one's life and lifestyle. It is too simplistic to say that the primary agents in these networks are corporations and non-profits working on behalf of and encouraged by a political rationality, particularly since this rationality stresses the importance of personal responsibility and enterprise. However it also is too simplistic to say, as I believe Jenkins does, that the participatory technologies available to TV viewers (particularly when aggregated into "grassroots communities") merely encourage or produce a more democratic citizenship than the old broadcast networks. Jenkins' perspective ignores the injunction to self-enterprise in the governmental rationality through which a "convergence culture" developed, and his perspective ignores the way that the virtue of free-choice, interactivity, and self-enterprise are objectified (made rational and a desired outcome) through citizens'/consumers' imbrication in the networks of government that television mediates. This section, therefore, considers the strategies of personalization and programmatization through which television (in its relation to these networks of government) has authorized a citizenship (one's relation to a governmental rationality and arrangement) that is free to choose, individualized, interactive, enterprising, and self-managed.

Foucault's reassessment of the liberal conception of *self-government* in terms of the history of the care of the self is instructive on this point. He argues that liberalism's valorization of individualism and freedom occurred through the multiplication of private authorities, institutions, programs, and technologies which objectify (make rational and knowable) and normalize various activities and behaviors. The technologies of government, which are dispersed through life in a particular time and place and which aim to make life healthy, not only work to discipline free individuals but assume (as Foucault emphasized in his latter work) a free subject with the capacity to act ethically—to govern her- or his-self properly. Beginning with the daily regimens of the Stoics, Foucault sketches the long history of "the care of the self," considering some of the ways that individuals have put and kept themselves on the right, healthy track through techniques for administering to oneself. Foucault follows the Stoic's practice of diary-keeping through Christian prayer and penitence, and subsequently through Modern scientific technologies of self-administration. With the birth of liberal government, becoming a subject who is self-governing thus involved mastering various "technologies of the self" and living one's life (everyday, every where) through a healthy regimen—the avoidance of excess and the measured, proper exercise of freedom.

Foucault's historical perspective about technologies of the self, self-writing, and care of the self is helpful in thinking about how media in the twentieth and twenty-first centuries operate in relation to the governmental rationalities and arrangements of liberalism, just as his historicism reminds us that liberalism can not be generalized (historically or geographically) and that it always is neces-sary to understand the liberal State and liberal government up through these technologies of self-government and citizenship. This perspective is particular-ly useful in thinking about the transition or conversion in the U.S. from the broadcasting technologies of citizenship (everyday programming for a nation-al audience) to the emergence of technologies of citizenship that developed in the current regime of media convergence and Reality TV and that have been marked by an economy of mass customization and interactivity. While it would be simplistic to paint this transformation in too broad strokes, suggesting that the practice of public-partnerships were unprecedented before the Reagan years and the ascendancy of cable and satellite era, that viewers of television during the broadcast era never "interacted" with television programming, or that cer-tain models and technologies of citizenship from the broadcast era are absent from the current rationality and (media) networks of government, there are important ways that the television-Web nexus has been organized to provide the resources for and encourage the formation of a citizenship that takes responsi-bility and shows enterprise in governing/caring for her/himself.

One vivid example of this trend is the ABC series, *Shaq's Big Challenge* (2006–2007). This series follows the quest of series co-producer and NBA basketball star, Shaquille O'Neal as he contemplates how to redress the "problem of child obesity." The series begins with O'Neal, as citizen, trying to figure out who can help him solve the problem. After weighing the resources available through "the halls of government," he eventually consults technical experts from the private sector, some of whom refer him to others. Galvanized by their knowledge about the problem, he then decides to assemble a "dream team" of experts (a fitness trainer, a child obesity specialist, and a professional nutritionist) to help him help seven children whom the series represents as "at risk" because of their size. The series title thus refers to the challenge of aiding these children (i.e., of governing their "unhealthy" conduct and physical size) but also to the challenge of managing the problem through the formation of a public-private network of authorities. The series charts the progress of both the kids and Shaq's team, as each episode sets the kids through a series of tests and contests aimed at instilling in them the confidence to help themselves. Like many Reality TV series that represent personal makeover as a "life intervention," this one deploys a mentor (in this case a celebrity citizen who relies on experts) literally to re-shape his subjects as self-directed, self-caring, and enterprising citizens. The series represents itself as bringing the children into a good and healthy network from which they have previously been excluded, in part (as pointed out in the first episode) because many of them have become obese/unhealthy watching too much of the wrong television, spending too much time on-line, or over-indulging in video games. In all these respects, *Shaq's Big Challenge* implicitly reflects on and represents itself as mediating a good and healthy media network.

Although the series mobilizes "private citizens" to provide a form of welfare, the series' references to the role of the State, corporate, and non-profit entities is significant. The corporate sponsors of the series' premier episode were predominantly interactive media companies (Verizon, Sprint, AT&T), private insurance providers (Gieco and Safe Auto), and pharmaceutical companies and brands (Pfizer, Celebrex, Lunesta), even though the children's treatment never explicitly mentions their need for a pharmacological regimen. As in *EM:HE*, these sponsors become part of the privatized network of welfare. At several key junctures in the series, Shaq makes reference to the challenge of convincing public school systems and State agencies in Florida (where the series is set) to become involved. In the first episode he twice refers to the importance of bringing his findings and his network to the attention of Florida's then-governor, Jeb Bush. Just as importantly, the series regularly invokes the President's Fitness Challenge and its criteria for healthy citizens as his primary yardstick for gauging the progress and health of his subjects. In this respect, the series goes further than most in formalizing the networks of the Bush administration's rationalization and authorization of public-private partnership–particularly significant since Florida's governor was the brother of the nation's president, and since the name of the TV series conjures the name of the State's fitness program. Yet the series also demonstrates that it is up to individual citizens to enact–or to find their proper place within–the proper *arrangement* of government for linking social enterprises of welfare with self-enterprise.

The series clearly works within the conventions of programs such as *The Biggest Loser* which designs televisual regimens (test and contests) for subjects deemed over-sized and unhealthy, and within the conventions of other ABC series whose life interventions mobilize private networks of public service. However, what makes this series different from many of the examples discussed in the prior section is its demonstration of civic involvement by a celebrity/exemplary citizen who reflects on the limits of his expertise in order to call forth a network of private practitioners—a facet of the series that makes it more like *Oprah's Big Give* than *Extreme Makeover: Home Edition*. Moreover Shaq's *private* odyssey, staged publically through this network of private practitioners, as well as Reality TV technicians, demonstrates a pathway that individuals can follow in order to become productive agents in a network of public service and welfare. In this sense, the series (at different levels, for different actor/agents) is about self-governing citizens and their networks. The kids must become enterprising under the guidance of Shaq's entrepreneurialism; ABC and the series' sponsors *act on* their enterprise, while federal and state government enter both as Shaq's supporters but also as institutions that he (the series and network) must mobilize into the proper arrangement (i.e., public-private partnership). In this arrangement, good government thus depends on the *actions* of citizens and the *support* the State. The citizen's self-awareness leads to producing a network of self-governing actors whose activity and network in turn help make the State aware of its resources.

More than many of the series discussed in the prior section, this one organizes the links between television programming and a network of Web-links in order to maximize this chain of enterprise and self-governing actors. While the series was on the air, its ABC website provided a special link to a related site, "Shaq's Big Family Challenge" (http://www.shaqsfamilychallenge.com/public-site/funnel/index.aspx). From that site, viewers could link to the .gov websites for Bush's "President's Fitness Challenge" (http://www.presidentschallenge.org/) and the President's Council on Fitness (http://www.fitness.gov/), the latter a site with multiple links from a menu (similar to ABC's Better Community site) of corporate and non-profit participants in the council. Just as importantly, the website provides various customizable resources for managing one's health *through* the series' TV-web nexus. Viewers who watch the program can become inter-/actively involved in managing a problem identified and overseen through this public-private network, tapping into toolkits and score-cards for making one's family, children, and self more accountable (or accountable within this rationality of government). The website for Shaq's Family Challenge, which has lingered after the cancelation of the TV series, provided a link in 2009 for the Everyday Health Network (http://www.everydayhealth.com/), a web-engine and company oriented entirely toward customizing a daily health regimen, finding experts who can address personal health problems, and connecting with non-profit and commercial "partners." In this respect, the series' website is like the self-managed websites for "lifestyle management" that insurance providers have increasingly encouraged for their policyholders in order to keep their customers less "at risk," as the bodies and accounts which these companies oversee: www.lifestylemanagement.healthlink.com. The website's linkage to an *everyday* "health network" (whose motto is "today's the day") not only moves the daili-ness of the TV regimen into the daily life and technical capacities of the inter-active subject, transforming the old idea of the "TV program" into one that is self-starting and self-directing, but it also technically mediates and helps pro-duce a network from television to the web resources for managing one's lifestyle.

To become fit, through Shaq's or Bush's "challenge," involves becoming an interactive, enterprising actor in this network—able to care for oneself, but through the technical resources necessary for full-fledged citizenship that the current complex of networks of government make possible. As a *self-authoriz-ing* network, *Shaq's Big Challenge* works to assemble multiple authorities (pub-lic and private) who assist and authorize a citizen-subject to take control of her or his own life—to become a freely operating authority about his or her self, in a network of provision that expects citizens to get fit and take care of themselves, and thus to break the shackles of their unhealthy dependence on the State.

Citizenship, as a technical achievement through these networks of govern-ment, not only entails becoming aware (arguably the old idea of citizenship in the broadcast era) but actively technologizing oneself through daily "challenges" and tests administered through networks of government that represent them-selves as part of everyday life and as arbiters/providers of private communities of citizens. It is no small coincidence that the websites for television networks in the first decade of the twenty-first century have adopted the generic link to "community," typically as part of a menu of link choices across the top of the website. The 2009 website for NBC's *The Biggest Loser* provided a Community-link that encourages its viewer to "stay motivated" by joining the Biggest Loser League—web-community as *team* of energized subjects whose game-playing (involvement in tests and contests) makes them healthier citizens. The Planet Green Network's website invites its visitors to perform their activism as green citizens by "joining the conversation" on the site, which provides links to Facebook and Twitter. As these examples suggest, the technical achievement of citizenship involves not only mastering the technologies of self-government, but in so doing entering the networks of privatized "community" and "public-private partnership." Membership in a Web-community attached to and authorized by a television channel involves procedures which replicate the achievement of citizenship, such as undergoing a registration process that will formalize a member's ability to vote and participate in various interactive polls, discussions, and games. However, the formalization of community membership (as privatized citizenship) is a technical prerequisite for recognition in the on-line community's membership as well as for participating in the forms of outreach and public service that are made available through the television-Web community. The fullness of this kind of citizenship depends on the extent one is willing to follow through the links sustained in what counts as the current everyday networks of government.

THE LEGACY OF THE PUBLIC-PRIVATE NETWORK, AND BARACK'S BIG CHALLENGE

The website for the recent Bravo and NBC series, *Parks & Recreation* (2009), a comedy of local government in a fictional Indiana city, provides interactive links to the program *and* to other commercial websites that attach their marketing campaign to NBC's. The series chronicles the efforts of Leslie Knope, the deputy director of the city's Parks and Recreation Department (played by Amy Poehler), as she navigates the procedures, policies, and networks of city government. While Knope is a paragon of the enterprising ethic operating within the halls of government, she never justifies her own behavior and agenda in quite the same way as one of her superiors who lectures her in the first episode about the virtue of public-private partnerships as a remedy to anything. In the first episode, Knope declares at a meeting with citizens that she intends to mobilize the resources of city government to build a public park on land abandoned by a private developer. Part of the series' humor results from the daffiness and even delusional quality of Knope's wide-eyed confidence that a public service can be so easily provided, especially when her confidence is played against the dysfunctionality of government administration and the rampant cynicism of citizens about whether government ever works for them. That the series derives humor from a deep cynicism about whether government ever works affirms a premise central to the recent liberal political rationality which called forth the public-private partnership, even as the series documents (a la Michael Moore) the bizarreness of that rationality and the failure of "reinvented government" to provide a substantively better alternative—particularly in a small city that increasingly is expected to provide services with limited State resources and to display an entrepreneurialism in lining up private, contracted providers. The rationality of recent liberal government's valorization of the public-private partnership assumes, after all, that local government knows best how to deliver social services, that the private sector can manage services better than can the State, and that citizens are freed from the "iron cage" of bureaucracy through the heroics and generosity of corporate and non-profit philanthropy.

There is therefore something completely familiar *and* ironic that the series' official website provides links to a website for the government offices of the fictional Pawnee. The home page of Pawnee's website follows the generic conventions of real city government websites: its heading is emblazoned with the motto, "My Hometown," underneath which is a menu of links for the city's various departments, such as Business, Public Safety, Information & Technology, Arts & Culture, Transportation, and Parks & Recreation. How should one consider the relation of Pawnee's website to those of gated communities whose private government administrations have had to develop policies and regulations about the disruptions attributed to the staging of Reality TV series in their community property?[6] Given the examples discussed above, there are reasons to expect that Pawnee may not be the last effort to reinvent the government of cities through the technical resources provided by television's current reinvention through the Web.

I mention this series as part of the Conclusion because its ambiguities bespeak a possible sea-change in the arrangement and rationality authorizing the everyday networks of government discussed in the prior sections. Although the history of "reinventing government" predates the Bush administration, it is no small coincidence that the networks of government attendant to the Reality TV syndrome flourished as part of the most energetically designed programs of federalism, public-private partnership, and self-enterprise during the Bush administration. So it is challenging to be assessing the legacy of this governmental rationality at a time when the election of Barack Obama and the financial crisis have been represented, on TV and other media, as the unraveling of this political rationality. The history gestured toward in this chapter is important in this respect because of several questions that it poses. Is there any indication that the conventions of the TV-Web nexus described herein have become fragile in 2009? Probably not. In that case, how persistent are the networks of government through and to which they have been activated and authorized? What for instance, are the traces of *Shaq's Big Challenge* on the website for the Obama administration's version of the President's Council on Fitness? Should we see a series which aired in the first months of the Obama administration, such as *HS, USA* or *The Wanted*, as anomalies in the current context (throwbacks to conventions and networks that are unraveling and no longer viable) or as a testament to the rootedness of this rationality in Obamaland? What has a "public plan" for healthcare (and accusations of its relation to a "new socialism") to do with "Barack's Big Challenge"—with the Obama Administration's provision of healthcare (and mobilizing healthcare reform) through the everyday networks and technologies of public-private partnership and of the care of the self that have formed during the previous eight years? What would it take, against the legacy of "reinventing government," to demonstrate the inefficacy, incompleteness, and inequalities of everyday networks of self-government that have sought to demonstrate and normalize their utility in empowering citizens to help themselves?

One of the signature events of the Obama administration's inauguration was the telecast, "We Are One: An Inaugural Celebration from the Lincoln Memorial," produced and distributed by HBO on January 18, 2009. Although the event tapped into the legacy of "grassroots" and "popular" demonstrations set since the 1960s along the Memorial's reflecting pool, this event's production and distribution by the largest pay-cable network attested to the importance of the public-private network of government from which the Obama administration *technically* was born. It served as a hallmark of a "post-broadcast era," when a national-popular civic event's "broadcast" occurred televisually through a pay-subscription service that allowed cable franchises to distribute the event publically "free of charge." HBO also made the event available through its web-site, hbo.com, at a time when HBO had begun extending some of its live programming (e.g., "Real Time with Bill Maher"—www.hbo.com/billmaher/) through on-line venues. HBO allowed National Public Radio to air the event, thus providing a relatively unprecedented linkage between PBS and TV subscription services. And HBO re-ran the event as part of its subscription-TV service for another month, when the telecast operated as a hybrid form of public philanthropy and programming for HBO's taste culture. Although the event was a celebration of political change—a display of citizenship and civic pride surrounding popular performers and entertainment companies performing without ostensible financial compensation—the meaning and mattering of "we are one" attest to both the flexibility and rootedness of the networks of government through which television is (re-)organized in a "post-broadcast era."

The flexibility and rootedness of these networks of government make it wrong to imagine that "Barack's Big Challenge" simply involves overcoming the Bush-Cheney Administration's promotion of an Ownership Society or that period's do-it-yourself technologies of Reality TV which supported those policies. In conjunction with having signed a "Serve America Act" (April 2009) and following its announcement (February 2009) that it would "expand" the Bush Administration's Faith-based & Community Initiative, the Obama

Administration began promoting its "United We Serve" program alongside week-long campaigns such as National Volunteer Week. The promotion and administration of the these programs occurred partly through the Obama Administration's United We Serve website (www.serve.gov and www.volunteer.gov), whose title banner states: "The President is calling on all Americans to participate in our nation's recovery and renewal by serving in our communities….America's new foundation will be built one community at a time—and it starts with you." Its claim of "renewal" stands in an important relation to "recovery," not simply as a broad, civic response to "economic recovery" but also to recovering and building on/through the established networks for administering welfare as public-private partnership. The National Volunteer Week's sponsors in April 2009 were both the United We Serve program and the corporate and non-profit coalition of the Points of Light Institute/Network, created under George H.W. Bush and merged in 2007 with the Hands On Network. While the Hands On Network website prominently lists a menu of its corporate sponsors, the United We Serve site lists a menu of state-based agencies and programs (e.g., AmeriCorp) that incentivize volunteerism, represent a point of interface between public and private "partnership," and provide technical resources for citizen participation. Furthermore, alongside this latter menu, The United We Serve website also prominently displays its network's connection to new media networks such as Facebook, Twitter, YouTube, and "Serve.gov Mobile" (the new presence of the state in the emergent complex of what Jenkins and others see as 'grassroots" media).

In conjunction and "partnership" with the birth and reinvention of these programs, EM:HE produced a broadcast whose introduction promoted itself as one staging area for National Volunteer Week—as part of a new season of the series which could hitch itself to Obama Administration initiatives as those initiatives acted on, renewed, and recovered the recent history of public-private partnership and volunteeristic programs comprising the residual networks of government. The episode of EM:HE thus rationalizes its latest outreach in terms of National Volunteer Week, even as it in turn contributes to a governmental rationality about the value of citizens (such as the Montgomery family showcased in the episode) who abandon "the fast-track" of personal financial gain in order to serve community through a life of volunteerism (for those disenfranchised from any 'track" through the financial crisis)—service hindered only, it would seem, by the lack of the technical resources linking that family and community (those citizens) to the more robust networks of government such as United We Serve and Thousand Points of Light. This renewing, renewable, recovering, and recoverable network of government (a synergy between the public and private mobilization of civic activism) acts on the past arrangement even as it rearticulates and reinvents the technical demonstrations and the reasoning about outreach and care.

CONCLUSION—"WE ARE ONE" (*OMNES ET SINGULATIM*) IN THE POPULAR RATIONALITIES & NETWORKS OF GOVERNMENT?

As noted in the Introduction to this chapter, Foucault explained the late eighteenth and early nineteenth century's birth of liberal government not simply in terms of the formation and concentration/expansion of the State's administrative apparatus but in terms of the "power of individualization" occurring through the proliferation and individualization of the technologies of control, discipline, and securitization. In his late lecture/essay, "Omnes et singulatim: Toward a Critique of Political Reason," he discussed this power of individualization (the government of the self) as having multiple administrators, including the relation of the individual to her/himself—the 'self' as an ethical substance which is worked on one's entire life. "Omnes" (the every-body) is thus managed through the work, care, and individualized power and technologies of the self.

Against this long (albeit Modern) history of liberal government, there is one other implication of "We Are One" that is worth considering, by way of a Conclusion (and an opening onto further questions and studies about "everyday networks of government" in the current historical conjuncture). This chapter's concern with television's place in everyday networks of government, and these networks' role in (re-)shaping forms of citizenship, focuses on how these networks are authorized through a political rationality of the "public-private partnership"—a reasoning about limiting the role of State government while catalyzing local (state and municipal) administration, corporate and non-profit institutions, and the enterprise and self-responsibilization of citizens. To the extent that the HBO production's title represents national collectivity as "*we* the people," and in that sense casts its production as a *public* good/service and a *popular* political representation, how should we reconcile that with the localization, privatization, and *individualization* of power and government valorized through the current ("neo-liberal") governmental rationality and the technological regime of the everyday networks of government which this chapter considers as having been renewed, reinvented, and/or recovered from the Bush Administration to the Obama Administration? Might a more useful way of thinking about the production's role in ceremonially/ritually marking the transformation of liberal government in the U.S. and in inaugurating President Obama's policy initiatives involve recognizing how the production *acts on* the profoundly individualized regime/economy (the everyday networks) of "popular rationalities" and "popular technologies" for administering government and citizenship—the collective "we" of a "popular culture" mobilized as/through the personalization, customization, and "interactivization" of collectivization and the "popular"? This is the technological regime through which broadcast media such as television (as "popular" media and "public" service) have been reinvented politically, commercially, and culturally.

Although *We are One* allows me to pose questions about what has and has not changed in the first year of the Obama Administration, it operates more within the ceremonial ritual of inauguration (the political protocols of national revival) than the everyday rituals of the television program, and the current televisual technologies of individuation and individualization. It is still, for all its implication in the rationality of the public-private partnership, a popular spectacle that rearticulates a national-popular history and civic-subjectivity before a mass audience—on TV and in front of the TV set. It acts upon the past from the current regime of interactivity—assembling the we, reforging a link between the nation and the people, in part by mixing old newsreel footage (e.g., old television scenes of the Martin Luther King rally at the Lincoln Memorial) with live musical performances demonstrating the potential to overcome the market and cultural differences of musical taste.

A new television series from 2009 that more strategically reconstitutes a collectivity through the power of *individualization* and through television's *everyday* networks of government is the ABC series *Find My Family*. Premiering in late 2009, the series deploys expert "researchers" to "sift through archives and track down records" in order to help individuals and individual families piece together old family ties. According to the series website: "We all know people who feel incomplete, searching for something to make them feel whole." Like *We are One*, *Find My Family* works to repair the We through the One (the individual family broken into pieces), but more than *We are One* it demonstrates as popular science that the hard work of family repair depends less on political leaders and Hollywood entertainers than 'teams" of technical experts (family repairmen and women) and the enterprise of individuals. As such, it operationalizes the current power of individualization—life lived and repaired through everyday networks of government, and through a form of televisuality that mediates government of and in daily life. Studying those everyday networks of government may not always or primarily lead to the linkages between State and private administrators/managers as tended to be emphasized in this chapter; sometimes it may involve following (to use Mattelart's expression) other "paths of reason" that coalesce or swarm around the everyday problem and object of administering government individually.

The examples of television's everyday networks of government offered in this chapter may not, for some readers, be the most vivid examples of the "global" or "post-global" orientation of these networks. For many reasons this chapters can only gesture toward how the analysis might move in that direction, though the chapter's consideration of *HS, USA* and *The Wanted* provide practical starting points for that discussion. However, this chapter provides ample justification for recognizing that a study of global governmentalities and the role of television in global networks of government should not fly too quickly over the mattering of the everyday formation of governmental rationalities, technologies, and networks. The problem with a study claiming to be global is just as problematic as one that generalizes liberalism in terms of the "neo," particularly a "neo-liberalism" that can claim to unfold in the same way everywhere. It is no small coincidence that one account of "neo-liberalism" developed through an account of global networks and economy (or the "network society"). While this chapter is not the venue to wade too deeply into an argument about theories and research of the global as neoliberal, the analytic that this chapter proposes more modestly begins where liberalism's work is never done—with the problems in managing the little, everyday government of life through the on-going (and increasingly individualized) technologies and networks of freedom. Accounting for the "global"- or "post-global"-ness of a "neo-liberal" governmental rationality begins by locating the changing/current techniques through which power and government are individualized in the everyday. To the extent that television still matters, it matters in part this way.

ACKNOWLEDGMENT

Some of the arguments in this chapter are ones that have benefited significantly from my conversations and writing collaborations with Laurie Ouellette, who has addressed similar points in her own writing. Some of the key points in section two of this chapter are ones that Laurie and I have developed collaboratively. I am, as always, deeply indebted to her help and insight.

NOTES

1. For an account of the political, economic, and cultural contradictions of PBS's formation, see Laurie Ouellette, *Viewers Like You?: How Public TV Failed the People*, New York: Columbia UP, 2002.
2. For annual amounts spent for contracting by the U.S. government, see link for Contracts at "USASpending.gov": http://www.usaspending.gov/fpds/index.php?reptype=a
3. For more on this series and its relation to ABC's Better Community initiative, see Chapter 1, "Charity TV", of Ouellette & Hay, 2008.
4. In 2008, the candidates for U.S. President (McCain, Clinton, and Obama) all taped videos supporting the initiative, though the videos never aired during the broadcasts.
5. See "Homeland Security Launches Expansion of Information Exchange System," February 2004, http://www.dhs.gov/xnews/releases/press_release_0354.shtm.
6. For examples of this, see Chapter 5, "TV's Constitutions of Citizenship, " Better Living through Reality TV, op cit.

BIBLIOGRAPHY

Andrejevic, Mark. (2003). *Reality TV: The Work of Being Watched*. New York: Rowman & Littlefield.

Castells, Manuel. (1996). *The Rise of the Network Society*. Oxford: Blackwell.

Deuze, Mark. (2007). *Media Work (Digital Media & Society)*. Cambridge: Polity Press.

Drucker, Peter. (1969). *The Age of Discontinuity*. New York: Harper & Row.

Foucault, Michel. (2001). *Power*. ed., James Faubion. New York: New Press.

Foucault, Michel. (2006). *Ethics: Subjectivity & Truth*. ed., Paul Rabinow. New York: New Press.

Foucault, Michel. (2008). *The Birth of Biopolitics: Lectures at the College de France (1978-1979)*. New York: Palgrave Macmillan.

Foucault, Michel. (2007/2009). *Security, Territory, Population: Lectures at the College de France (1977-1978)*. New York: Palgrave Macmillan/Picador.

Hartley, John. (2007). *Television Truths: Forms of Knowledge in Popular Culture*. Boston: Wiley-Blackwell.

Jenkins, Henry. (2006). *Convergence Culture: Where Old & New Media Collide*. New York: NYU Press.

Lotz, Amanda. (2007). *The Television will he Revolutionized*. New York: NYU Press.

Mattelart, Armand. (1996). *The Invention of Communication*. Minneapolis: University of Minnesota Press.

Negroponte, Nicholas. (1996). *Being Digital*. New York: Vintage Books/Random House.

Osborne, David & Gabler, Ted. (1992). *Reinventing Government: How the Entrepreneurial Spirit is Transforming the Public Sector*. Reading, MA: Addison-Wesley.

Ouellette, Laurie. (2002). *Viewers Like You?: How Public TV Failed the People*. New York: Columbia University Press.

Ouellette, Laurie & Hay, James. (2008). *Better Living through Reality TV*. Boston:Wiley-Blackwell.

Rose, Nikolas. (1999). *Powers of Freedom: Reframing Political Thought*. Cambridge/New York: Cambridge University Press.

Von Hayek, Friedrich. (1944). *The Road to Serfdom*. Chicago: University of Chicago Press.

Von Hayek, Friedrich. (1960). *The Constitution of Liberty*. Chicago: University of Chicago Press.

CHAPTER TEN

Privacy As Work

The Appropriation of Labor in Post-Global Network

GRANT KIEN

The hybrid socio-technological evolution within our post-global network necessitates a method of capitalist appropriation of labor that has yet to be fully schematized theoretically. Seminal works by authors such as Mosco (1998), Castells (2000, 2001), Hardt and Negri (2000), Stratton (2000) and others have normalized network as our contemporary global model of both production and consumption. In terms of production, distributed models of development and manufacturing are so commonplace that terms such as "just in time" delivery and "outsourcing," popular just a decade ago, sound old fashioned. In terms of consumption, globalization of consumer culture and cultural imperialist projects (Cvetkovich and Kellner, 1997; Xie, 2008) have normalized "America" as the slick facade of global culture.

Studies of labor migration and the evolution of teleworkers, net-workers, and the like have helped us understand the bio-machinic nature of sweat-shops, such as in the case of Chinese gold farmers, and in other spaces where-in labor erupts in the global network.[1] The "Cybertariat" (Huws, 2001) and "Binary Proletariat" (Bolt, 2000) are concepts already well-constructed and nor-malized in discourse. Traditional pre-global network formations of capitalism have thus been identified within the recently developed context of globalized production and consumption that have been enabled by global network. Important as such theorizations have been to the evolution of thinking through labor in the context of globalization and "network power," these tools of analysis are not sufficient for our present and future, that is, for post-globally net-worked sites of production and consumption. This is not to say that the older formations of production and consumption have all disappeared. It is to say, rather, that a new form has erupted amid the old.

We have entered a new era of labor. This new era calls for a new theoreti-cal tool to understand how capitalism appropriates labor even while such labor-ing entails an act of consumption. Capitalist agents achieve this through a techno-evolutionary re-appropriation of what we have traditionally thought of as privacy. One might argue there is nothing genuinely new about this emer-gent condition. Capitalism first created and then has steadily commodified leisure time throughout its evolution. A new spike in capitalism, however, is apace. What has begun to change since the 1980s and has recently exploded with the Web 2.0 social networking storm that began approximately 2007 is the very category of participation in the commodification of leisure. The leisure consumer has, in effect, become a leisure-laborer, who produces his or her own commodi-ties along with and as part of the consumption of spectacle. This type of pro-duction doesn't happen alongside or in dialectical relation to consumption. It is not a simple blurring of the line between production and consumption; it is instead complexly bound up with the consumerist act itself.

In its simplest explanation, the act of consumption creates data (Castells 2000, 2001; Mosco, 1998). Data is, of course, the principal commodity in an information economy. While this concept is perhaps somewhat basic and rela-tively easy to understand, the major change I'm referring to is not. The techno-evolutionary change I am interested in charting indexes the relation between the assemblage known as the global media apparatus and those consumer-pro-duced commodities (i.e., user generated content) upon which the media appa-ratus depends in order to maintain its stability and perpetually feed itself. This position proceeds from the premise that we are living in a condition of post-global digital consumer culture (Kien, forthcoming), in which consumption manifests as commodified "authentic" experiences delivered through mediation. An important part of this is the fulfillment of Alvin Toffler's (1980) vision of the "prosumer," the consumer that participates in the production of what they consume. [2] The following case study, focusing on the production and reproduc-tion of visual artwork, demonstrates how this works on a global scale. These works of art function as a model through which to explain the post-globally net-worked media apparatus which ubiquitously assimilates us.

THE PROSUMER

In 1980, futurist Alvin Toffler put forward the preposterous notion that one day consumers would participate in the production of what they consume. He called this subject the "Prosumer." Thirty years later, it is safe to say that we have clear examples of Toffler's projected subject. The futurist's dream becomes a present reality through numerous consumerist norms that have been established in the past decade due to the proliferation of consumerist access to technology and a resulting propensity for customization of products. For example, Nike online, Build a Baby and its counterpart Build a Bear, and even Burger King all focus on the selling of self-chosen, customized, and therefore "original" products which the consumer assembles by selecting features from a range of options. Consumer choices in product interface (aka "preferences") and content are a central consideration in digital products, including social networking sites such as Facebook and MySpace. However, even while exponentially proliferating our range of choice, the age of the prosumer reveals the fundamental problem with limits and regulations. The problem is that one size does not indeed fit all. In fact, one size does not actually perfectly fit anyone. Or at least not anyone we really know. Attempts at regulation, instead, function within a diligent appropriation of labor designed to feed a system of production and consumption. There can never be enough choice to satisfy everyone. Furthermore, practically speaking, the more choice the consumer is confronted with, the more choices are needed to produce the temporally based illusion of authenticity of experience.

The impermanence of subjective identity is precisely the issue here. Susan Lee Star (1991) deftly illustrated how the "uniqueness" of identity can cause a halt in the otherwise smoothly running function of systems due to an inability of machinic systems to adapt in the face of unique orders falling outside their programmatic rationale. In a slightly different way, Janice Radway (in Hay, 1996) pointed out that the audience is not a fixed, unchanging identity spanning consecutive moments. Rather, media is perpetually confronted by identity "shapeshifters" who are one thing one moment but who transform into something completely different the next, i.e., the manager who in the next moment becomes a mother, only to become a consumer in the next moment. This is not to say that stability in identity doesn't exist. It is to say, however, that a stable identity is so rare that it is freakish. This is our state of normalcy as audience members and consumers in the post-global network. The identity of pirate/thief has quickly become normalized as one of our commonly inhabited roles in new media environments. Prosumerism brings with it numerous adjunct identities.

Who is not a pirate in some capacity these days? I am not merely referencing downloaded movies and music. I am also invoking more ubiquitous and banal recirculations of images and information on blogs and through email. Who is not a pornographer, eroticizing the public sphere? Who is not a voyeur, spying on a world filled with strangers and friends alike with applications like Twitter and YouTube postings? At the same time, who is not somehow whoring him or herself: is anyone exempt from the habit of trading one's own intimacies in exchange for something considered valuable, e.g., online profiles? Who is not taking advantage of disparities in the global economic system, either overtly or covertly, perhaps even unknowingly seeking competitive advantages on a global scale? What happens to notions of responsibility in this context? This identity shape-shifting and normalization of piracy, voyeurism, pornography, global exploitation and even quasi-prostitution are key aspects that make us participants in the techno-evolution of privacy.

I don't want to romanticize the shift that is taking place, but rather to focus on the hidden appropriations of labor that are happening in all these above situations as we progress into a post-globally networked world. On the one hand, many people enjoy a great deal of flexibility and convenience as a result of the technological world we're creating. On the other hand, however, not all of this is desirable. Old world hierarchies exist disguised in new world formats. Marshall McLuhan (1964) illustrated how the content of new media is precisely the old media it replaces. In much the same way oppressions and appropriations of labor remain in post-global network; they are simply remediated into new formats. Network and virtual spaces provide new territory for old world hierarchies such as racism (Nakamura, 2001), sexism (Wajcman, 1991), ableism[3] and other forms of discrimination to manifest. Class and labor are likewise forms of exploitation remediated in the context of global network. Spotting new manifestations of old hierarchies in the new media environment can be tricky. The clarity of the phenomenon can be doubly confounded by the extent to which we in the guise of prosumer subjects participate in our own exploitation. To demonstrate how this works in the context of post-global network, I'll turn briefly to the artwork of Brendan Lott.

GLOBAL CAPITALISM AS ART: THE CASE OF BRENDAN LOTT'S NON-MEMORY

In the summer of 2008, I encountered a collection of photorealist oil paintings in a small installation room at the back of the San Jose Institute of Contemporary Art in California. The art institute was exhibiting the canvasses of Brendan Lott, who draped his unusual, and even bizarre, subjects in the convention of soft porn.[4] His models were portrayed as transgressive, kinky and deviant. At the same time they were recognizable as everyday, "real" people. While Lott presented the viewer with powerful images in terms of subject matter, the viewer is more likely to be impressed by the strange "everydayness" of his photorealistic images.

Brendan Lott, *Duplex Odalisque*, Oil on Canvas, 22" x 28", 2007,
Courtesy of Baer Ridgway Exhibitions

Adding to this feeling of everyday intimacy, Lott's paintings have a flat, snapshot-like quality to them. They refuse the artful depth of composition we might expect from a master painter. Why would this be? Is Lott trying to ride the wave of the reality TV aesthetic currently popular in contemporary production? Is he trying to reproduce the popularity of amateur composition dominating global outlets such as YouTube? These seem like viable answers if one conceives Lott's paintings in terms of traditional derivative reproduction, in which an artist acquires an image and fashions it into a new form for consumption. However, Lott is not an artist in this traditional sense. In fact, Lott didn't capture the images in his paintings himself. Upon investigation, one finds he didn't even paint his own canvases.

Lott culls the images in his paintings from the internet. He selects them from the plethora of user-posted images available for download over P2P (peer to peer) file sharing communities such as Gnutella and BitTorrent. After acquiring the images, Lott sends the image files to an oil-painting service in China, where they are reproduced by hand on canvas. Upon receipt, Lott prepares and mounts the painted canvases as his installation.

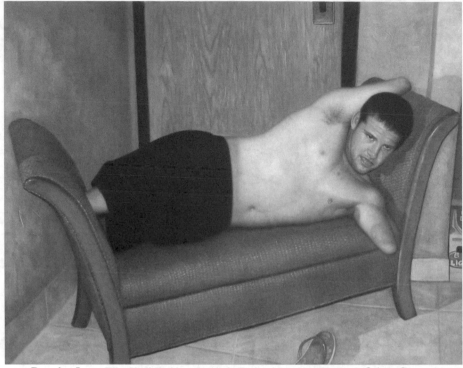

Brendan Lott, *Who Shall Fashion Anew the Body of Our Humiliation?*, Oil on Canvas,
20" x 27", 2007, Courtesy of Baer Ridgway Exhibitions

Brendan Lott takes advantage of our post-global situation to appropriate both content and labor in the production of his artwork. He quite effectively demonstrates a complete co-option of labor with virtual tools which is made possible by the advantages of his privileged position as an American in a global context. His work is enabled by an appropriation of images,[5] appropriation of creative labor that produces such content, and an appropriation of global manufacturing labor, i.e., 3rd world oppression. His end product subversively exposes the institutional-commercial art system by confounding Geertz's (1973) "authenticity" machine. Yet how is this possible? How do we understand this process to work in the present global context? Reviewing the phenomenon of self-pornografication through a couple of popular Web 2.0 sites will help illustrate the conditions that make Lott's appropriation work as art.

The web site Wretch.cc was started several years ago in Taiwan by a couple of college boys who began posting pictures of girls whom they found attractive. They then opened up the site to girls to self-post their own images. Wretch quickly became filled with self-pornographying images posted by thousands of self-promoting young women and girls. It was eventually opened up to males as well, who also began posting self-pornographying images. A similar idea in

Brendan Lott, *As a Dog Returns to Its Vomit, So a Fool Repeats His Folly*, Oil on Canvas, 27" x 36", 2008, Courtesy of Baer Ridgway Exhibitions

video, Stickam.com, is a live, streaming video chat site that attracts a mainly teen clientele. When one goes inside the site, it's obvious there is again a phenomenon of self-pornogrification occurring. I reference Wretch and Stickam as exemplary sites from which images may be poached, as they provide an ever expanding and inexhaustible supply of sensationalist user-created content. The bulk of the work—creating the images—has already been done for the poacher. All that remains is to pluck the raw content from its original context and then feed it into the manufacturing machine to emerge as a finished product. While Lott works with the more anonymous P2P networks, he could just as easily glean his content from commercial sites like those described above, and which are meaningful for the discussion below of how the corporate machine feeds itself. Nonetheless, Lott is purposeful in his harvesting, in order to create an artful intervention. In the process he illustrates a way of appropriating content that is very much akin to the way the contemporary global media machine works. His work serves as an easily understood case study that exemplifies the way I endeavor to explain the converged global media system as it now functions. Let's now look at the larger media system and the way it appropriates labor as content.

POST-GLOBALLY NETWORKED MEDIA AND THE APPROPRIATION OF PRIVACY

The 2008 Eliot Spitzer scandal[6] reached a critical point with a text message: "Pls let me know if [Client-9's] 'package' arrives 2mrw. Appt wd b on Wed."[7] "Client-9" is none other than scandalized politician Eliot Spitzer. As an early pioneer in electronic archiving policy[8], it's remarkable that Spitzer left such an electronic trail in the first place. But more relevant to this discussion is that the general surveillance of Spitzer begun with an FBI wiretap and investigation didn't end with his press conference. After a certain tipping point, the FBI's involvement was barely needed in order to keep tabs on Spitzer's whereabouts and moment to moment activities. The sensationalism of his case aside, the Spitzer scandal demonstrates the way the contemporary media apparatus works; that is, the media apparatus is a ubiquitous, 24/7 surveillance machine. The major network news followed Spitzer's motorcade through downtown New York to his live press conference, turning even the act of riding in a limousine into media content. But it's not just the major media networks that participate in this harvesting of content. A sophisticated, profoundly interconnected cross-platform/cross media system of production works to supply various points of consumption for the end-users (aka audience members).

Private citizens working with YouTube, passive surveillance systems in public spaces, insiders "leaking" information to various media outlets, virtual piles of electronic information left in our wakes as we go through our days—these all became sources of content for the media apparatus, just as P2P and image sharing web sites provide content for smaller scale appropriations exemplified above by Brendan Lott. Theorists beginning with Debord and the Situationists pointed out that the society of spectacle needs these spectacular events as a vital component of our economy. The information machine that constitutes the fabric of our system moves from one event like this to another in a constant cycle of consumption in its purest form, the consumption of unrefined messages. This is content for the sake of the medium.

This is not to say that there isn't some substance to the charges against Spitzer or to diminish the meaning of the ensuing scandal. But as media consumers we see this script enacted over and over again. The "reality" TV sting in particular allows the audience to participate as surveilling sleuths. The audience is encouraged to add content through web site interfaces and to become involved in the outcomes of specific cases and events. This windfall of free content allows networks to poach, enrich and prolong stories far beyond what used to be considered a natural news cycle in the glory days of dinner time network TV. Thus, when Spitzer himself was no longer forthcoming as a story, the apparatus simply sustained the story of his misdeed through its own apparatus of artificial life support. Attention moved to the prostitute, Ashley Dupre, who happened to have a MySpace page dedicated to her "real" career as a singer/performer. This conveniently provided some enticing content for viewers: excellent melodrama written by Dupre herself. Dupre's is the story of a bad girl from an abusive, broken home trying to get her life straight.[9] From a public relations perspective, she couldn't have asked for a better opportunity to manage her media presence.

Unavailable for comment, either by choice, or by legal gag order, Dupre's MySpace presence became her subjectivity, reported on by all the major news outlets and quoted extensively in the *New York Times*. Those inclined to actually visit her MySpace page find out Dupre is not alone. In fact, she had an outpouring of support that can be read at the bottom of her page. For example, take this comment from Dupre's old classmate:

> MarLo-CONFIDENZIALE
> Mar 14 2008 8:49 AM
>
> Ashley remember what we learned in our media class...free publicity is THE BEST.
> Roll with the punches and keep your head up :)[10]

Within two days of the scandal breaking in the media, hundreds of messages were sent to her "friend's comments" section. But who are these friends? It turns out that MarLo-CONFIDENZIALE had been busy constructing her own MySpace page promoting her career services, which, as par for course, was linked from her comment to Dupre. This is the principle of viral marketing in action, considered by many in the industry to be one of the most powerful aspects of the social networking phenomenon.

Not all such viral marketers are as active in their messages. Some don't even bother writing a message. Rather, they just post their link without the labor of writing anything more. This form of marketing is already well-established. I bring it up here because it demonstrates the way the post-global network media machine operates. It is content for the sake of the medium. Like an unquantifiably large, self-perpetuating feedback loop, the media apparatus feeds on this kind of spectacle even while it feeds into its very production. At the same time, all new inputs add critical mass to the system (Virilio, 2000) that takes on exponential growth through the harvesting, recycling and re-appropriation of all relative content. It cannot stop, lest the whole network collapse under its own weight.

A few more examples effectively illustrate the point: the Mark Foley intern scandal,[11] the Tom Cruise "Scientology rant,"[12] the Hong Kong movie star sex scandal involving Edison Chen and his numerous famous lovers[13] all became points of focus inspiring the harvest of raw media material which directly involve some sort of cross platform/mixed media assemblage. Everyday life has become like a 24/7 reality TV show. But it's not just TV, it's reality media. It's not just the condition of media stars; it's a fact of everyday life. This is true for almost everyone in our post-global networked condition. Simply regard "Joe the Plumber" who capitalized on it during the 2008 presidential campaign, and reveals on his own web site:

> Since the debate 10–15–08 between John McCain and Barrack Obama, I have received: 363,000 hits on my website, over 500 phone calls, and 600 e-mails. I have been interviewed for several publications and appeared on Television 5 times, interviewed for the Radio 6 times, made the cover of the Local newspaper twice and met a lot of great people, many of which flattered me by asking for autographs.[14]

The ubiquity of our current, everyday 24/7 reality media show belies the myth of privacy. There used to be some method of shutting out the mediascape, i.e., pulling the plug. Now even when you shut off your appliances, they're still recording your habits—pulling the plug doesn't matter, as there is no "plug" anymore. Our virtual selves never stop working. Bank accounts, blogs, investment programs, mapping systems…Google, Yahoo, Microsoft and other companies record it all. Our TIVO and cable boxes note what we watch, when we quit and start viewing, and the hours in between. Our web browsers, cell phones, GPS units[15], etc., all do the same, noting our media consumption choices and habits. Even the photos we upload to our personal computers are subject to being cataloged by cookies from web photo services. Every detail of electronic life is logged and immortalized in digital format. This is why identity theft is so devastating; once it's logged, it's extremely difficult, if not impossible, to erase.

The vitality of junk information is incredible.[16] It circulates for years. There are still junk e-mails from the early years of the internet circulating from user to user. Once digital information is released into the techno-ether, there's no way to recall it. While you may request removal from their cache archiving system, Google and other companies make copies of everything they find on-line and unless requested to delete specific pages, store it in perpetuity. This constitutes only part of that which we are publically allowed to know. Most of the citizenry is in fact quite unaware of the extent to which military and covert agency surveillance software abounds, as do keystroke loggers, snooping programs, hijack programs etc., which can be verified by simply visiting warez (aka Hacker/cracker) sites. After the events of 9/11, the FBI publically announced it was activating its Carnivore email snooping program to surveil the internet.[17] China has more publically proven it is possible to strictly, and effectively, monitor the internet and other digital communications media that form the network. Among other countries, Australia openly expressed interest in these Chinese surveillance programs. As a result, privacy no longer exists the way it once did, as a retreat. Our modernist notion of privacy as escape from the world no longer seems tenable. As Donald Kerr, former deputy director of U.S. National Intelligence put it, "Too often, privacy has been equated with anonymity; and it's an idea that is deeply rooted in American culture." (Kerr, 2007) Many people might bristle at this suggestion, but Kerr is facing an uncomfortable fact that many are not yet prepared to confront.

In a "heterotopic" (Foucault, 1967) condition created by global network, there is no longer a useful distinction between public and private space (Kien, 2009). The combination of telepresence (Virilio, 2000) with global hybrid network that seamlessly interweaves virtual and physical environments (Kien, in press) results in the conceptual management of all space and time. In this condition we have virtual subjectivities that work for us 24/7. Equally, there are virtual subjectivities and "bots" out there working against you 24/7. The imperative is clear: if YOU don't manage your digital world, someone else will. The task of managing identity in the present context is to attempt to manage all time and space, which is, of course, impossible. Thus the unavoidable proliferation of content from our everyday mediated lives, whether intentionally produced or not, sits open and vulnerable to harvest from the global media apparatus. This apparatus is invested in culling from the lives of all, from major celebrity to commoner alike, in order to appropriate private labor in both the production and consumption of media products.

The Appropriation of Labor in Post-Global Network

To return to a meta-vision of the topic at hand, Brendan Lott's installation *Memories I'll Never Have* demonstrates on a micro scale the way the global media system appropriates images, content, and creative labor. It also demonstrates how global labor fulfills a networked logic of manufacturing even while reproducing traditional hierarchy, i.e., 3rd world oppression. In the process, he subverts the art system by manufacturing a crisis in the notion of authenticity. Lott positions himself within the machine. Entrenched and dependent on it, he demonstrates how Web 2.0 P2P sites garnish a volunteering of labor while at the same time he appropriates user-generated content. The phenomenon of prosumerism then disseminates a redistribution of that content for global consumption.

Meanwhile, the case of Spitzer and other celebrity scandals illustrate our new definition of privacy in which the media machine constantly scours the global network for content to feed itself. The operation of the media machine highlights an emergent need for a knowingly self-managed media presence. To reiterate the major points of this essay: there can be no turning back from this situation, as the apparatus has no off switch. It is a seemingly perpetual motion machine in which every media act (both production and consumption) feeds itself. Indeed, information itself becomes the commodity (Castells, 2001) which we readily produce as part and parcel of our consumerism. This is a new form of capitalist appropriation of labor that goes beyond self-branding, beyond identity maintenance. It is the result of a lifelong campaign of identity management, trapping the post-globally networked individual in an everyday labor relation not of his or her choosing.

Notes

1. See http://www.chinesegoldfarmers.com/ and http://www.nytimes.com/2007/06/17/magazine/17lootfarmers-t.html.
2. The term "prosumer" as I use it here is not to be confused with the "professional consumer" class of consumer goods.
3. See http://ableismonline.wordpress.com
4. See http://www.abdelmortt.com/
5. It bears mentioning that Lott's artwork and the sites discussed below rely on a continuous sexualization and commodification of women's bodies as their primary aesthetic commodity. This foregrounds a much needed feminist critique that I look forward to engaging as this discourse moves forward.
6. See http://www.nytimes.com/2008/03/10/nyregion/10cnd-spitzer.html
7. See http://www.nydailynews.com/news/2008/03/10/2008-03-10_client_9_will_pay_for_everything.html
8. In terms of e-mail archiving in particular, Spitzer, above others, should have been fully aware of the electronic trail he was leaving and how it might be used against him. See http://www.ferris.com/?p=319680
9. See http://www.myspace.com/ninavenetta
10. See http://www.myspace.com/marlo-confidenziale
11. See http://en.wikipedia.org/wiki/Mark_Foley: http://seattlepi.nwsource.com/local/223201_west06.html,http://www.planetout.com/pno/news/article.html?date=2003/12/01/1
12. See http://www.youtube.com/watch?v=UFBZ_uAbxS0 and http://www.youtube.com/watch?v=H-C-wupe76E&feature=related
13. See http://news.about-knowledge.com/edison-chen-scandal-photos/ and http://en.wikipedia.org/wiki/Edison_Chen.
14. Retrieved from http://www.joelaratheplumber.com/ on 17 Aug., 2009.
15. GPS units effectively turn the act of driving into an act of media consumption, all the while recording one's driving habits. This in turn transforms driving into an act of media production.
16. See http://www.snopes.com/ and http://www.truthorfiction.com/.
17. The program was then later abandoned and replaced with a commercially available snooping program. See http://www.foxnews.com/story/0,2933,144809,00.html.

REFERENCES

Bolt, N. The binary proletariat. *First Monday.* 2000, v5, n5. Online: http://firstmonday.org/issues/issue5_5/bolt/index.html

Castells, M. (2000). *The Rise of the Network Society,* 2nd edition. Malden: Blackwell.

———. (2001). *The Internet Galaxy: Reflections on the Internet, Business, and Society.* New York: Oxford University Press.

Cvetkovich, A., and Kellner, D. (Eds.), (1997). Articulating the global and the local: *Globalization and Cultural Studies.* Boulder: Westview Press, pp. 1–32.

Foucault, M. (1967). Of other spaces: Heterotopias, translated by Jay Miskowiec from the original "Des Espace Autres" in Architecture/Mouvement/ContinuitŽ, October 1984. Accessed online January 2, 2005: http://foucault.info/documents/heteroTopia/foucault.heteroTopia.en.html

Geertz, C. (1973). *The Interpretation of Cultures.* New York: Basic Books.

Hardt, M., and Negri, A. (2000). *Empire.* Cambridge, MA: Harvard University Press.

Hay, J. (1996). Afterword. In J. Hay, L. Grossberg and E. Wartella (Eds.), *The Audience and it's Landscape.* Boulder: Westview Press.

Huws, U. The making of a cybertariat: Virtual work in a real world. *Socialist Register.* 2001, 1–23.

Kien, G. (2009). *Global Technography: Ethnography in the Age of Mobility.* New York: Peter Lang.

———. (Forthcoming). *Digital Consumer Culture.* Manuscript in production.

———. (in press). Virtual environment: The machine is our world. In S. Tettegah & C. Cologne (Eds.), *Identity, Learning and Support in Virtual Environment.* Netherlands: Sense Publishers.

McLuhan, M. (1964). *Understanding Media.* Mentor: Penguin.

Mosco, V. (1998). *The Political Economy of Communication.* Thousand Oaks: Sage Publications.

Nakamura, L. (2001). Race in/for cyberspace: Identity tourism and racial passing on the Internet. In D. Trend (Ed.), *Reading Digital Culture.* Malden: Blackwell.

Star, S. L. (1991). Power, technologies and the phenomenology of conventions: On being allergic to onions. In J. Law (Ed.), *A Sociology of Monsters: Essays on Power, Technology and Domination, Sociological Review Monograph 38.* London: Routledge, pp. 26–56.

Stratton, J. (2000). Cyberspace and the globalization of culture. In D. Bell and B. Kennedy (Eds.), *The Cybercultures Reader.* New York: Routledge, pp. 721–731.

Toffler, A. (1980). *The Third Wave.* New York: William Morrow & Co.

Virilio, P. (2000). *The Information Bomb.* Chris Turner (trans). New York: Verso.

Wajcman, J. (1991). *Feminism Confronts Technology.* Cambridge: Blackwell Polity.

Xie, S. (2008). Anxieties of modernity: A semiotic analysis of globalization images in China. *Semiotica,* 170–1/4, pp. 153–168.

About the Authors

Anca N. Birzescu is a doctoral candidate in communication studies at Bowling Green State University, Ohio. Her research interests include postcolonial theory, feminist cultural studies, ethnic identity, gender and communication, and mass media discourse in Eastern and Central European transitional societies. Her current doctoral research focuses on Roma minority identity negotiation in post-communist Romania.

Jack Zeljko Bratich, PhD, is an associate professor of journalism and media studies at Rutgers University. He authored *Conspiracy Panics: Political Rationality and Popular Culture* (2008) and coedited, along with Jeremy Packer and Cameron McCarthy, *Foucault, Cultural Studies, and Governmentality* (2003). His work applies autonomist social theory to such topics as reality television, social media, and the cultural politics of secrecy. He is currently writing a book titled *Programming Reality* (under contract with Lexington), which examines reality programs (on and off television) as experiments in affective convergence.

Radhika Gajjala, PhD, is professor of communication and cultural studies, and Director of Women's Studies at Bowling Green State University, Ohio. Her book *Cyberselves: Feminist Ethnographies of South Asian Women* (2004) was published by Altamira Press. She co-edited *South Asian Technospaces* and *Webbing Cyberfeminist Practice* and is currently working on a single-authored book entitled *Technocultural Agency: Production of Identity at the Interface* (under contract with Lexington). Url: http://personal.bgsu.edu/~radhik

Michael D. Giardina, PhD, is visiting assistant professor of advertising and affiliate faculty of cultural studies & interpretive research at the University of Illinois at Urbana-Champaign. He is the author or editor of nine books, including *Sporting Pedagogies: Performing Culture & Identity in the Global Arena* (Peter Lang, 2005), *Youth Culture & Sport: Identity, Power, & Politics* (with Michele K. Donnelly, Routledge, 2007), and *Contesting Empire/Globalizing Dissent: Cultural Studies after 9/11* (with Norman K. Denzin, Paradigm, 2006). He is also the associate editor of the *Sociology of Sport Journal* and a member of the editorial board of *Cultural Studies/Critical Methodologies*. With Joshua I. Newman, he is completing a book titled *Consuming NASCAR Nation: Sport, Spectacle, and the Politics of Neoliberalism*.

Ulrike Gretzel, PhD, is an assistant professor in recreation, park and tourism sciences at Texas A&M University and director of the Laboratory for Intelligent Systems in Tourism. Her research focuses on persuasion in the context of human-computer interaction, the mediation of tourism experiences by emerging mobile and web-based technologies, and the representation of sensory experiences and emotions through digital media.

James Hay, PhD, is an associate professor in the Institute of Communications Research and the department of media & cinema studies at the University of Illinois at Urbana-Champaign. He is the co-author of *Better Living through Reality TV* (Blackwell, 2008).

Grant Kien, PhD, is an assistant professor with the department of communication at California State University, East Bay. His research focuses on technography, qualitative approaches to technology research, globalization, communication and culture, mobility and communications networks as performative, symbolic, and interpretive spaces. Recent works include the full length book entitled *Global Technography: Ethnography in the Age of Mobility* (Peter Lang, 2009), a chapter in the book *Material Culture and Technology in Everyday Life: Ethnographic Approaches* (edited by Phillip Vannini, Peter Lang, 2009), and guest editor of a partial special issue on technology and qualitative research in the scholarly journal *Qualitative Inquiry* (Sage, 2008).

Marina Levina, PhD, is a faculty member in the media studies program at UC Berkeley. She is currently working on a book titled *Life as a Virus, Life as a Code: Biopolitics of Control over Post-Human Life*. She has published work on personal genomics, genetic engineering, and cultural metaphors of scientific research. Her research interests include critical studies of science and technology, visual culture, and critical cultural theory.

Joy Pierce, PhD, is assistant professor in the department of communication at the University of Utah. Her work focuses on information technology and marginalized populations. Dr. Pierce employs critical cultural and social theories as well as qualitative research methods to explore issues of inclusion and exclusion from computer use and new media consumerism to new technology research and development. Her works in new and emerging technologies, information communication policy, social problems and feminist multicultural theory have been presented in national and international conferences. She has served as guest editor for a special issue in *Social Identities*, and has published in *Television and New Media*, *Studies in Symbolic Interaction*, and *Minorities and Communication*. Dr. Pierce began her career in print journalism. She teaches writing for new media, communication technology and culture, qualitative research methods, as well as socio-cultural, applications-based courses and seminars in emerging technologies.

James Salvo is a PhD candidate in the Institute of Communications Research at the University of Illinois at Urbana-Champaign. By day, he researches new media, psychoanalytic theory, and book culture, but by night, he's asleep.

Sean Smith is an artist and critical sport theorist living in Toronto. He is currently a doctoral candidate in the media and communications program at the European Graduate School and a lecturer at various universities in Canada. He also holds degrees in sport management from Acadia University and the International Institute for the Study of Sport Management at the University of Alberta. sportsBabel is his weblog of research notes and other works (www.sportsbabel.net).

Index